S0-BMZ-830

The Sciences of the Artificial

The Sciences of the Artificial

Third edition

Herbert A. Simon

The MIT Press
Cambridge, Massachusetts
London, England

This book was set in Sabon by Graphic Composition, Inc.

Printed and bound in the United States of America.

Library of Congress Cataloging-in-Publication Data

Simon, Herbert Alexander, 1916–
    The sciences of the artificial / Herbert A. Simon.—3rd ed.
      p.   cm.
    Includes bibliographical references and index.
    ISBN 0-262-19374-4 (alk. paper).—ISBN 0-262-69191-4 (pbk.: alk. paper)
    1. Science—Philosophy.   I. Title.
Q175.S564   1996
300.1'1—dc20   96-12633
                                                                    CIP

To Allen Newell
in memory of a friendship

Contents

Preface to Third Edition

As the Earth has made more than 5,000 rotations since *The Sciences of the Artificial* was last revised, in 1981, it is time to ask what changes in our understanding of the world call for changes in the text.

Of particular relevance is the recent vigorous eruption of interest in complexity and complex systems. In the previous editions of this book I commented only briefly on the relation between general ideas about complexity and the particular hierarchic form of complexity with which the book is chiefly concerned. I now introduce a new chapter to remedy this deficit. It will appear that the devotees of complexity (among whom I count myself) are a rather motley crew, not at all unified in our views on reductionism. Various among us favor quite different tools for analyzing complexity and speak nowadays of "chaos," "adaptive systems," and "genetic algorithms." In the new chapter 7, "Alternative Views of Complexity" ("The Architecture of Complexity" having become chapter 8), I sort out these themes and draw out the implications of artificiality and hierarchy for complexity.

Most of the remaining changes in this third edition aim at updating the text. In particular, I have taken account of important advances that have been made since 1981 in cognitive psychology (chapters 3 and 4) and the science of design (chapters 5 and 6). It is gratifying that continuing rapid progress in both of these domains has called for numerous new references that record the advances, while at the same time confirm and extend the book's basic theses about the artificial sciences. Changes in emphases in chapter 2 reflect progress in my thinking about the respective roles of organizations and markets in economic systems.

This edition, like its predecessors, is dedicated to my friend of half a lifetime, Allen Newell—but now, alas, to his memory. His final book, *Unified Theories of Cognition,* provides a powerful agenda for advancing our understanding of intelligent systems.

I am grateful to my assistant, Janet Hilf, both for protecting the time I have needed to carry out this revision and for assisting in innumerable ways in getting the manuscript ready for publication. At the MIT Press, Deborah Cantor-Adams applied a discerning editorial pencil to the manuscript and made communication with the Press a pleasant part of the process. To her, also, I am very grateful.

In addition to those others whose help, counsel, and friendship I acknowledged in the preface to the earlier editions, I want to single out some colleagues whose ideas have been especially relevant to the new themes treated here. These include Anders Ericsson, with whom I explored the theory and practice of protocol analysis; Pat Langley, Gary Bradshaw, and Jan Zytkow, my co-investigators of the processes of scientific discovery; Yuichiro Anzai, Fernand Gobet, Yumi Iwasaki, Deepak Kulkarni, Jill Larkin, Jean-Louis Le Moigne, Anthony Leonardo, Yulin Qin, Howard Richman, Weimin Shen, Jim Staszewski, Hermina Tabachneck, Guojung Zhang, and Xinming Zhu. In truth, I don't know where to end the list or how to avoid serious gaps in it, so I will simply express my deep thanks to all of my friends and collaborators, both the mentioned and the unmentioned.

In the first chapter I propose that the goal of science is to make the wonderful and the complex understandable and simple—but not less wonderful. I will be pleased if readers find that I have achieved a bit of that in this third edition of *The Sciences of the Artificial.*

Herbert A. Simon
Pittsburgh, Pennsylvania
January 1, 1996

Preface to Second Edition

This work takes the shape of a fugue, whose subject and countersubject were first uttered in lectures on the opposite sides of a continent and the two ends of a decade but are now woven together as the alternating chapters of the whole.

The invitation to deliver the Karl Taylor Compton lectures at the Massachusetts Institute of Technology in the spring of 1968 provided me with a welcome opportunity to make explicit and to develop at some length a thesis that has been central to much of my research, at first in organization theory, later in economics and management science, and most recently in psychology.

In 1980 another invitation, this one to deliver the H. Rowan Gaither lectures at the University of California, Berkeley, permitted me to amend and expand this thesis and to apply it to several additional fields.

The thesis is that certain phenomena are "artificial" in a very specific sense: they are as they are only because of a system's being molded, by goals or purposes, to the environment in which it lives. If natural phenomena have an air of "necessity" about them in their subservience to natural law, artificial phenomena have an air of "contingency" in their malleability by environment.

The contingency of artificial phenomena has always created doubts as to whether they fall properly within the compass of science. Sometimes these doubts refer to the goal-directed character of artificial systems and the consequent difficulty of disentangling prescription from description. This seems to me not to be the real difficulty. The genuine problem is to show how empirical propositions can be made at all about systems that, given different circumstances, might be quite other than they are.

Almost as soon as I began research on administrative organizations, some forty years ago, I encountered the problem of artificiality in almost its pure form:

. . . administration is not unlike play-acting. The task of the good actor is to know and play his role, although different roles may differ greatly in content. The effectiveness of the performance will depend on the effectiveness of the play and the effectiveness with which it is played. The effectiveness of the administrative process will vary with the effectiveness of the organization and the effectiveness with which its members play their parts. [*Administrative Behavior,* p. 252]

How then could one construct a theory of administration that would contain more than the normative rules of good acting? In particular, how could one construct an empirical theory? My writing on administration, particularly in *Administrative Behavior* and part IV of *Models of Man,* has sought to answer those questions by showing that the empirical content of the phenomena, the necessity that rises above the contingencies, stems from the inabilities of the behavioral system to adapt perfectly to its environment—from the limits of rationality, as I have called them.

As research took me into other areas, it became evident that the problem of artificiality was not peculiar to administration and organizations but that it infected a far wider range of subjects. Economics, since it postulated rationality in economic man, made him the supremely skillful actor, whose behavior could reveal something of the requirements the environment placed on him but nothing about his own cognitive make-up. But the difficulty must then extend beyond economics into all those parts of psychology concerned with rational behavior—thinking, problem solving, learning.

Finally, I thought I began to see in the problem of artificiality an explanation of the difficulty that has been experienced in filling engineering and other professions with empirical and theoretical substance distinct from the substance of their supporting sciences. Engineering, medicine, business, architecture, and painting are concerned not with the necessary but with the contingent—not with how things are but with how they might be—in short, with design. The possibility of creating a science or sciences of design is exactly as great as the possibility of creating any science of the artificial. The two possibilities stand or fall together.

These essays then attempt to explain how a science of the artificial is possible and to illustrate its nature. I have taken as my main examples the fields of economics (chapter 2), the psychology of cognition (chapters 3 and 4), and planning and engineering design (chapters 5 and 6). Since Karl Compton was a distinguished engineering educator as well as a distinguished scientist, I thought it not inappropriate to apply my conclusions about design to the question of reconstructing the engineering curriculum (chapter 5). Similarly Rowan Gaither's strong interest in the uses of systems analysis in public policy formation is reflected especially in chapter 6.

The reader will discover in the course of the discussion that artificiality is interesting principally when it concerns complex systems that live in complex environments. The topics of artificiality and complexity are inextricably interwoven. For this reason I have included in this volume (chapter 8) an earlier essay, "The Architecture of Complexity," which develops at length some ideas about complexity that I could touch on only briefly in my lectures. The essay appeared originally in the December 1962 *Proceedings of the American Philosophical Society.*

I have tried to acknowledge some specific debts to others in footnotes at appropriate points in the text. I owe a much more general debt to Allen Newell, whose partner I have been in a very large part of my work for more than two decades and to whom I have dedicated this volume. If there are parts of my thesis with which he disagrees, they are probably wrong; but he cannot evade a major share of responsibility for the rest.

Many ideas, particularly in the third and fourth chapters had their origins in work that my late colleague, Lee W. Gregg, and I did together; and other colleagues, as well as numerous present and former graduate students, have left their fingerprints on various pages of the text. Among the latter I want to mention specifically L. Stephen Coles, Edward A. Feigenbaum, John Grason, Pat Langley, Robert K. Lindsay, David Neves, Ross Quillian, Laurent Siklóssy, Donald S. Williams, and Thomas G. Williams, whose work is particularly relevant to the topics discussed here.

Previous versions of chapter 8 incorporated valuable suggestions and data contributed by George W. Corner, Richard H. Meier, John R. Platt, Andrew Schoene, Warren Weaver, and William Wise.

A large part of the psychological research reported in this book was supported by the Public Health Service Research Grant MH-07722 from the National Institute of Mental Health, and some of the research on design reported in the fifth and sixth chapters, by the Advanced Research Projects Agency of the Office of the Secretary of Defense (SD-146). These grants, as well as support from the Carnegie Corporation, the Ford Foundation, and the Alfred P. Sloan Foundation, have enabled us at Carnegie-Mellon to pursue for over two decades a many-pronged exploration aimed at deepening our understanding of artificial phenomena.

Finally, I am grateful to the Massachusetts Institute of Technology and to the University of California, Berkeley, for the opportunity to prepare and present these lectures and for the occasion to become better acquainted with the research in the sciences of the artificial going forward on these two stimulating campuses.

I want to thank both institutions also for agreeing to the publication of these lectures in this unified form, The Compton lectures comprise chapters 1, 3, and 5, and the Gaither lectures, chapters 2, 4, and 6. Since the first edition of this book (The MIT Press, 1969) has been well received, I have limited the changes in chapters 1, 3, 5, and 8 to the correction of blatant errors, the updating of a few facts, and the addition of some transitional paragraphs.

The Sciences of the Artificial

1

Understanding the Natural and the Artificial Worlds

About three centuries after Newton we are thoroughly familiar with the concept of natural science—most unequivocally with physical and biological science. A natural science is a body of knowledge about some class of things—objects or phenomena—in the world: about the characteristics and properties that they have; about how they behave and interact with each other.

The central task of a natural science is to make the wonderful commonplace: to show that complexity, correctly viewed, is only a mask for simplicity; to find pattern hidden in apparent chaos. The early Dutch physicist Simon Stevin, showed by an elegant drawing (figure 1) that the law of the inclined plane follows in "self-evident fashion" from the impossibility of perpetual motion, for experience and reason tell us that the chain of balls in the figure would rotate neither to right nor to left but would remain at rest. (Since rotation changes nothing in the figure, if the chain moved at all, it would move perpetually.) Since the pendant part of the chain hangs symmetrically, we can snip it off without disturbing the equilibrium. But now the balls on the long side of the plane balance those on the shorter, steeper side, and their relative numbers are in inverse ratio to the sines of the angles at which the planes are inclined.

Stevin was so pleased with his construction that he incorporated it into a vignette, inscribing above it

Wonder, en is gheen wonder

that is to say: "Wonderful, but not incomprehensible."

This is the task of natural science: to show that the wonderful is not incomprehensible, to show how it can be comprehended—but not to

Figure 1
The vignette devised by Simon Stevin to illustrate his derivation of the law of the inclined plane

destroy wonder. For when we have explained the wonderful, unmasked the hidden pattern, a new wonder arises at how complexity was woven out of simplicity. The aesthetics of natural science and mathematics is at one with the aesthetics of music and painting—both inhere in the discovery of a partially concealed pattern.

The world we live in today is much more a man-made,[1] or artificial, world than it is a natural world. Almost every element in our environment shows evidence of human artifice. The temperature in which we spend most of our hours is kept artificially at 20 degrees Celsius; the humidity is added to or taken from the air we breathe; and the impurities we inhale are largely produced (and filtered) by man.

Moreover for most of us—the white-collared ones—the significant part of the environment consists mostly of strings of artifacts called "symbols" that we receive through eyes and ears in the form of written and spoken language and that we pour out into the environment—as I am now doing—by mouth or hand. The laws that govern these strings of

1. I will occasionally use "man" as an androgynous noun, encompassing both sexes, and "he," "his," and "him" as androgynous pronouns including women and men equally in their scope.

symbols, the laws that govern the occasions on which we emit and receive them, the determinants of their content are all consequences of our collective artifice.

One may object that I exaggerate the artificiality of our world. Man must obey the law of gravity as surely as does a stone, and as a living organism man must depend for food, and in many other ways, on the world of biological phenomena. I shall plead guilty to overstatement, while protesting that the exaggeration is slight. To say that an astronaut, or even an airplane pilot, is obeying the law of gravity, hence is a perfectly natural phenomenon, is true, but its truth calls for some sophistication in what we mean by "obeying" a natural law. Aristotle did not think it natural for heavy things to rise or light ones to fall (*Physics,* Book IV); but presumably we have a deeper understanding of "natural" than he did.

So too we must be careful about equating "biological" with "natural." A forest may be a phenomenon of nature; a farm certainly is not. The very species upon which we depend for our food—our corn and our cattle—are artifacts of our ingenuity. A plowed field is no more part of nature than an asphalted street—and no less.

These examples set the terms of our problem, for those things we call artifacts are not apart from nature. They have no dispensation to ignore or violate natural law. At the same time they are adapted to human goals and purposes. They are what they are in order to satisfy our desire to fly or to eat well. As our aims change, so too do our artifacts—and vice versa.

If science is to encompass these objects and phenomena in which human purpose as well as natural law are embodied, it must have means for relating these two disparate components. The character of these means and their implications for certain areas of knowledge—economics, psychology, and design in particular—are the central concern of this book.

The Artificial

Natural science is knowledge about natural objects and phenomena. We ask whether there cannot also be "artificial" science—knowledge about artificial objects and phenomena. Unfortunately the term "artificial" has a pejorative air about it that we must dispel before we can proceed.

My dictionary defines "artificial" as, "Produced by art rather than by nature; not genuine or natural; affected; not pertaining to the essence of the matter." It proposes, as synonyms: affected, factitious, manufactured, pretended, sham, simulated, spurious, trumped up, unnatural. As antonyms, it lists: actual, genuine, honest, natural, real, truthful, unaffected. Our language seems to reflect man's deep distrust of his own products. I shall not try to assess the validity of that evaluation or explore its possible psychological roots. But you will have to understand me as using "artificial" in as neutral a sense as possible, as meaning man-made as opposed to natural.[2]

In some contexts we make a distinction between "artificial" and "synthetic." For example, a gem made of glass colored to resemble sapphire would be called artificial, while a man-made gem chemically indistinguishable from sapphire would be called synthetic. A similar distinction is often made between "artificial" and "synthetic" rubber. Thus some artificial things are imitations of things in nature, and the imitation may use either the same basic materials as those in the natural object or quite different materials.

As soon as we introduce "synthesis" as well as "artifice," we enter the realm of engineering. For "synthetic" is often used in the broader sense of "designed" or "composed." We speak of engineering as concerned with "synthesis," while science is concerned with "analysis." Synthetic or artificial objects—and more specifically prospective artificial objects having desired properties—are the central objective of engineering activity and skill. The engineer, and more generally the designer, is concerned with how things *ought* to be—how they ought to be in order to *attain goals,*

2. I shall disclaim responsibility for this particular choice of terms. The phrase "artificial intelligence," which led me to it, was coined, I think, right on the Charles River, at MIT. Our own research group at Rand and Carnegie Mellon University have preferred phrases like "complex information processing" and "simulation of cognitive processes." But then we run into new terminological difficulties, for the dictionary also says that "to simulate" means "to assume or have the mere appearance or form of, without the reality; imitate; counterfeit; pretend." At any rate, "artificial intelligence" seems to be here to stay, and it may prove easier to cleanse the phrase than to dispense with it. In time it will become sufficiently idiomatic that it will no longer be the target of cheap rhetoric.

and to *function*. Hence a science of the artificial will be closely akin to a science of engineering—but very different, as we shall see in my fifth chapter, from what goes currently by the name of "engineering science."

With goals and "oughts" we also introduce into the picture the dichotomy between normative and descriptive. Natural science has found a way to exclude the normative and to concern itself solely with how things are. Can or should we maintain this exclusion when we move from natural to artificial phenomena, from analysis to synthesis?[3]

We have now identified four indicia that distinguish the artificial from the natural; hence we can set the boundaries for sciences of the artificial:

1. Artificial things are synthesized (though not always or usually with full forethought) by human beings.

2. Artificial things may imitate appearances in natural things while lacking, in one or many respects, the reality of the latter.

3. Artificial things can be characterized in terms of functions, goals, adaptation.

4. Artificial things are often discussed, particularly when they are being designed, in terms of imperatives as well as descriptives.

The Environment as Mold

Let us look a little more closely at the functional or purposeful aspect of artificial things. Fulfillment of purpose or adaptation to a goal involves a relation among three terms: the purpose or goal, the character of the artifact, and the environment in which the artifact performs. When we think of a clock, for example, in terms of purpose we may use the child's definition: "a clock is to tell time." When we focus our attention on the clock itself, we may describe it in terms of arrangements of gears and the

---

3. This issue will also be discussed at length in my fifth chapter. In order not to keep readers in suspense, I may say that I hold to the pristine empiricist's position of the irreducibility of "ought" to "is," as in chapter 3 of my *Administrative Behavior* (New York: Macmillan, 1976). This position is entirely consistent with treating natural or artificial goal-seeking systems as phenomena, without commitment to their goals. *Ibid.*, appendix. See also the well-known paper by A. Rosenbluth, N. Wiener, and J. Bigelow, "Behavior, Purpose, and Teleology," *Philosophy of Science, 10* (1943):18–24.

application of the forces of springs or gravity operating on a weight or pendulum.

But we may also consider clocks in relation to the environment in which they are to be used. Sundials perform as clocks *in sunny climates*—they are more useful in Phoenix than in Boston and of no use at all during the Arctic winter. Devising a clock that would tell time on a rolling and pitching ship, with sufficient accuracy to determine longitude, was one of the great adventures of eighteenth-century science and technology. To perform in this difficult environment, the clock had to be endowed with many delicate properties, some of them largely or totally irrelevant to the performance of a landlubber's clock.

Natural science impinges on an artifact through two of the three terms of the relation that characterizes it: the structure of the artifact itself and the environment in which it performs. Whether a clock will in fact tell time depends on its internal construction and where it is placed. Whether a knife will cut depends on the material of its blade and the hardness of the substance to which it is applied.

The Artifact as "Interface"

We can view the matter quite symmetrically. An artifact can be thought of as a meeting point—an "interface" in today's terms—between an "inner" environment, the substance and organization of the artifact itself, and an "outer" environment, the surroundings in which it operates. If the inner environment is appropriate to the outer environment, or vice versa, the artifact will serve its intended purpose. Thus, if the clock is immune to buffeting, it will serve as a ship's chronometer. (And conversely, if it isn't, we may salvage it by mounting it on the mantel at home.)

Notice that this way of viewing artifacts applies equally well to many things that are not man-made—to all things in fact that can be regarded as adapted to some situation; and in particular it applies to the living systems that have evolved through the forces of organic evolution. A theory of the airplane draws on natural science for an explanation of its inner environment (the power plant, for example), its outer environment (the character of the atmosphere at different altitudes), and the relation between its inner and outer environments (the movement of an airfoil through a gas). But a theory of the bird can be divided up in exactly the same way.[4]

Given an airplane, or *given* a bird, we can analyze them by the methods of natural science without any particular attention to purpose or adaptation, without reference to the interface between what I have called the inner and outer environments. After all, their behavior is governed by natural law just as fully as the behavior of anything else (or at least we all believe this about the airplane, and most of us believe it about the bird).

Functional Explanation

On the other hand, if the division between inner and outer environment is not necessary to the analysis of an airplane or a bird, it turns out at least to be highly convenient. There are several reasons for this, which will become evident from examples.

Many animals in the Arctic have white fur. We usually explain this by saying that white is the best color for the Arctic environment, for white creatures escape detection more easily than do others. This is not of course a natural science explanation; it is an explanation by reference to purpose or function. It simply says that these are the kinds of creatures that will "work," that is, survive, in this kind of environment. To turn the statement into an explanation, we must add to it a notion of natural selection, or some equivalent mechanism.

An important fact about this kind of explanation is that it demands an understanding mainly of the outer environment. Looking at our snowy surroundings, we can predict the predominant color of the creatures we are likely to encounter; we need know little about the biology of the creatures themselves, beyond the facts that they are often mutually hostile, use visual clues to guide their behavior, and are adaptive (through selection or some other mechanism).

4. A generalization of the argument made here for the separability of "outer" from "inner" environment shows that we should expect to find this separability, to a greater or lesser degree, in *all* large and complex systems, whether they are artificial or natural. In its generalized form it is an argument that all nature will be organized in "levels." My essay "The Architecture of Complexity," included in this volume as chapter 8, develops the more general argument in some detail.

Analogous to the role played by natural selection in evolutionary biology is the role played by rationality in the sciences of human behavior. If we know of a business organization only that it is a profit-maximizing system, we can often predict how its behavior will change if we change its environment—how it will alter its prices if a sales tax is levied on its products. We can sometimes make this prediction—and economists do make it repeatedly—without detailed assumptions about the adaptive mechanism, the decision-making apparatus that constitutes the inner environment of the business firm.

Thus the first advantage of dividing outer from inner environment in studying an adaptive or artificial system is that we can often predict behavior from knowledge of the system's goals and its outer environment, with only minimal assumptions about the inner environment. An instant corollary is that we often find quite different inner environments accomplishing identical or similar goals in identical or similar outer environments—airplanes and birds, dolphins and tunafish, weight-driven clocks and battery-driven clocks, electrical relays and transistors.

There is often a corresponding advantage in the division from the standpoint of the inner environment. In very many cases whether a particular system will achieve a particular goal or adaptation depends on only a few characteristics of the outer environment and not at all on the detail of that environment. Biologists are familiar with this property of adaptive systems under the label of homeostasis. It is an important property of most good designs, whether biological or artifactual. In one way or another the designer insulates the inner system from the environment, so that an invariant relation is maintained between inner system and goal, independent of variations over a wide range in most parameters that characterize the outer environment. The ship's chronometer reacts to the pitching of the ship only in the negative sense of maintaining an invariant relation of the hands on its dial to the real time, independently of the ship's motions.

Quasi independence from the outer environment may be maintained by various forms of passive insulation, by reactive negative feedback (the most frequently discussed form of insulation), by predictive adaptation, or by various combinations of these.

Functional Description and Synthesis

In the best of all possible worlds—at least for a designer—we might even hope to combine the two sets of advantages we have described that derive from factoring an adaptive system into goals, outer environment, and inner environment. We might hope to be able to characterize the main properties of the system and its behavior without elaborating the detail of *either* the outer or inner environments. We might look toward a science of the artificial that would depend on the relative simplicity of the interface as its primary source of abstraction and generality.

Consider the design of a physical device to serve as a counter. If we want the device to be able to count up to one thousand, say, it must be capable of assuming any one of at least a thousand states, of maintaining itself in any given state, and of shifting from any state to the "next" state. There are dozens of different inner environments that might be used (and have been used) for such a device. A wheel notched at each twenty minutes of arc, and with a ratchet device to turn and hold it, would do the trick. So would a string of ten electrical switches properly connected to represent binary numbers. Today instead of switches we are likely to use transistors or other solid-state devices.[5]

Our counter would be activated by some kind of pulse, mechanical or electrical, as appropriate, from the outer environment. But by building an appropriate transducer between the two environments, the physical character of the interior pulse could again be made independent of the physical character of the exterior pulse—the counter could be made to count anything.

Description of an artifice in terms of its organization and functioning—its interface between inner and outer environments—is a major objective of invention and design activity. Engineers will find familiar the language of the following claim quoted from a 1919 patent on an improved motor controller:

What I claim as new and desire to secure by Letters Patent is:
1 In a motor controller, in combination, reversing means, normally effective field-weakening means and means associated with said reversing means for

5. The theory of functional equivalence of computing machines has had considerable development in recent years. See Marvin L. Minsky, *Computation: Finite and Infinite Machines* (Englewood Cliffs, N.J.: Prentice-Hall, 1967), chapters 1–4.

rendering said field-weakening means ineffective during motor starting and thereafter effective to different degrees determinable by the setting of said reversing means . . . [6]

Apart from the fact that we know the invention relates to control of an electric motor, there is almost no reference here to specific, concrete objects or phenomena. There is reference rather to "reversing means" and "field-weakening means," whose further purpose is made clear in a paragraph preceding the patent claims:

The advantages of the special type of motor illustrated and the control thereof will be readily understood by those skilled in the art. Among such advantages may be mentioned the provision of a high starting torque and the provision for quick reversals of the motor.[7]

Now let us suppose that the motor in question is incorporated in a planing machine (see figure 2). The inventor describes its behavior thus:

Referring now to [figure 2], the controller is illustrated in outline connection with a planer (100) operated by a motor M, the controller being adapted to govern the motor M and to be automatically operated by the reciprocating bed (101) of the planer. The master shaft of the controller is provided with a lever (102) connected by a link (103) to a lever (104) mounted upon the planer frame and projecting into the path of lugs (105) and (106) on the planer bed. As will be understood, the arrangement is such that reverse movements of the planer bed will, through the connections described, throw the master shaft of the controller back and forth between its extreme positions and in consequence effect selective operation of the reversing switches (1) and (2) and automatic operation of the other switches in the manner above set forth.[8]

In this manner the properties with which the inner environment has been endowed are placed at the service of the goals in the context of the outer environment. The motor will reverse periodically under the control of the position of the planer bed. The "shape" of its behavior—the time path, say, of a variable associated with the motor—will be a function of the "shape" of the external environment—the distance, in this case, between the lugs on the planer bed.

The device we have just described illustrates in microcosm the nature of artifacts. Central to their description are the goals that link the inner

6. U.S. Patent 1,307,836, granted to Arthur Simon, June 24, 1919.
7. Ibid.
8. Ibid.

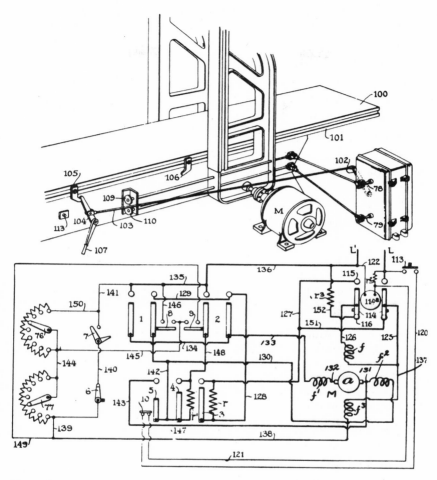

Figure 2
Illustrations from a patent for a motor controller

to the outer system. The inner system is an organization of natural phenomena capable of attaining the goals in some range of environments, but ordinarily there will be many functionally equivalent natural systems capable of doing this.

The outer environment determines the conditions for goal attainment. If the inner system is properly designed, it will be adapted to the outer environment, so that its behavior will be determined in large part by the behavior of the latter, exactly as in the case of "economic man." To predict how it will behave, we need only ask, "How would a rationally designed system behave under these circumstances?" The behavior takes on the shape of the task environment.[9]

Limits of Adaptation

But matters must be just a little more complicated than this account suggests. "If wishes were horses, all beggars would ride." And if we could always specify a protean inner system that would take on exactly the shape of the task environment, designing would be synonymous with wishing. "Means for scratching diamonds" defines a design objective, an objective that *might* be attained with the use of many different substances. But the design has not been achieved until we have discovered at least one realizable inner system obeying the ordinary natural laws—one material, in this case, hard enough to scratch diamonds.

Often we shall have to be satisfied with meeting the design objectives only approximately. Then the properties of the inner system will "show through." That is, the behavior of the system will only partly respond to the task environment; partly, it will respond to the limiting properties of the inner system.

Thus the motor controls described earlier are aimed at providing for "quick" reversal of the motor. But the motor must obey electromagnetic and mechanical laws, and we could easily confront the system with a task where the environment called for quicker reversal than the motor was capable of. In a benign environment we would learn from the motor only what it had been called upon to do; in a taxing environment we would learn something about its internal structure—specifically about those aspects of the internal structure that were chiefly instrumental in limiting performance.[10]

---

9. On the crucial role of adaptation or rationality—and their limits—for economics and organization theory, see the introduction to part IV, "Rationality and Administrative Decision Making," of my *Models of Man* (New York: Wiley, 1957); pp. 38–41, 80–81, and 240–244 of *Administrative Behavior*; and chapter 2 of this book.

10. Compare the corresponding proposition on the design of administrative organizations: "Rationality, then, does not determine behavior. Within the area of rationality behavior is perfectly flexible and adaptable to abilities, goals, and

A bridge, under its usual conditions of service, behaves simply as a relatively smooth level surface on which vehicles can move. Only when it has been overloaded do we learn the physical properties of the materials from which it is built.

Understanding by Simulating

Artificiality connotes perceptual similarity but essential difference, resemblance from without rather than within. In the terms of the previous section we may say that the artificial object imitates the real by turning the same face to the outer system, by adapting, relative to the same goals, to comparable ranges of external tasks. Imitation is possible because distinct physical systems can be organized to exhibit nearly identical behavior. The damped spring and the damped circuit obey the same second-order linear differential equation; hence we may use either one to imitate the other.

Techniques of Simulation

Because of its abstract character and its symbol manipulating generality, the digital computer has greatly extended the range of systems whose behavior can be imitated. Generally we now call the imitation "simulation," and we try to understand the imitated system by testing the simulation in a variety of simulated, or imitated, environments.

Simulation, as a technique for achieving understanding and predicting the behavior of systems, predates of course the digital computer. The model basin and the wind tunnel are valued means for studying the behavior of large systems by modeling them in the small, and it is quite certain that Ohm's law was suggested to its discoverer by its analogy with simple hydraulic phenomena.

---

knowledge. Instead, behavior is determined by the irrational and nonrational elements that bound the area of rationality . . . administrative theory must be concerned with the limits of rationality, and the manner in which organization affects these limits for the person making a decision." *Administrative Behavior,* p. 241. For a discussion of the same issue as it arises in psychology, see my "Cognitive Architectures and Rational Analysis: Comment," in Kurt VanLehn (ed.), *Architectures for Intelligence* (Hillsdale, NJ: Erlbaum, 1991).

Simulation may even take the form of a thought experiment, never actually implemented dynamically. One of my vivid memories of the Great Depression is of a large multicolored chart in my father's study that represented a hydraulic model of an economic system (with different fluids for money and goods). The chart was devised by a technocratically inclined engineer named Dahlberg. The model never got beyond the pen-and-paint stage at that time, but it could be used to trace through the imputed consequences of particular economic measures or events—provided the theory was right![11]

As my formal education in economics progressed, I acquired a disdain for that naive simulation, only to discover after World War II that a distinguished economist, Professor A. W. Phillips had actually built the Moniac, a hydraulic model that simulated a Keynesian economy.[12] Of course Professor Phillips's simulation incorporated a more nearly correct theory than the earlier one and was actually constructed and operated—two points in its favor. However, the Moniac, while useful as a teaching tool, told us nothing that could not be extracted readily from simple mathematical versions of Keynesian theory and was soon priced out of the market by the growing number of computer simulations of the economy.

Simulation as a Source of New Knowledge

This brings me to the crucial question about simulation: *How can a simulation ever tell us anything that we do not already know?* The usual implication of the question is that it can't. As a matter of fact, there is an interesting parallelism, which I shall exploit presently, between two assertions about computers and simulation that one hears frequently:

1. A simulation is no better than the assumptions built into it.
2. A computer can do only what it is programmed to do.

I shall not deny either assertion, for both seem to me to be true. But despite both assertions simulation can tell us things we do not already know.

11. For some published versions of this model, see A. O. Dahlberg, *National Income Visualized* (N.Y.: Columbia University Press, 1956).

12. A. W. Phillips, "Mechanical Models in Economic Dynamics," *Economica*, New Series, *17* (1950):283–305.

There are two related ways in which simulation can provide new knowledge—one of them obvious, the other perhaps a bit subtle. The obvious point is that, even when we have correct premises, it may be very difficult to discover what they imply. All correct reasoning is a grand system of tautologies, but only God can make direct use of that fact. The rest of us must painstakingly and fallibly tease out the consequences of our assumptions.

Thus we might expect simulation to be a powerful technique for deriving, from our knowledge of the mechanisms governing the behavior of gases, a theory of the weather and a means of weather prediction. Indeed, as many people are aware, attempts have been under way for some years to apply this technique. Greatly oversimplified, the idea is that we already know the correct basic assumptions, the local atmospheric equations, but we need the computer to work out the implications of the interactions of vast numbers of variables starting from complicated initial conditions. This is simply an extrapolation to the scale of modern computers of the idea we use when we solve two simultaneous equations by algebra.

This approach to simulation has numerous applications to engineering design. For it is typical of many kinds of design problems that the inner system consists of components whose fundamental laws of behavior—mechanical, electrical, or chemical—are well known. The difficulty of the design problem often resides in predicting how an assemblage of such components will behave.

Simulation of Poorly Understood Systems

The more interesting and subtle question is whether simulation can be of any help to us when we do not know very much initially about the natural laws that govern the behavior of the inner system. Let me show why this question must also be answered in the affirmative.

First, I shall make a preliminary comment that simplifies matters: we are seldom interested in explaining or predicting phenomena in all their particularity; we are usually interested only in a few properties abstracted from the complex reality. Thus, a NASA-launched satellite is surely an artificial object, but we usually do not think of it as "simulating" the moon or a planet. It simply obeys the same laws of physics, which relate only to its inertial and gravitational mass, abstracted from most of its other properties. It *is* a moon. Similarly electric energy that entered my house from the early atomic generating station at Shippingport did not "simulate" energy generated by means of a coal plant or a windmill. Maxwell's equations hold for both.

The more we are willing to abstract from the detail of a set of phenomena, the easier it becomes to simulate the phenomena. Moreover we do not have to know, or guess at, all the internal structure of the system but only that part of it that is crucial to the abstraction.

It is fortunate that this is so, for if it were not, the topdown strategy that built the natural sciences over the past three centuries would have been infeasible. We knew a great deal about the gross physical and chemical behavior of matter before we had a knowledge of molecules, a great deal about molecular chemistry before we had an atomic theory, and a great deal about atoms before we had any theory of elementary particles—if indeed we have such a theory today.

This skyhook-skyscraper construction of science from the roof down to the yet unconstructed foundations was possible because the behavior of the system at each level depended on only a very approximate, simplified, abstracted characterization of the system at the level next beneath.[13] This is lucky, else the safety of bridges and airplanes might depend on the correctness of the "Eightfold Way" of looking at elementary particles.

Artificial systems and adaptive systems have properties that make them particularly susceptible to simulation via simplified models. The characterization of such systems in the previous section of this chapter

13. This point is developed more fully in "The Architecture of Complexity," chapter 8 in this volume. More than fifty years ago, Bertrand Russell made the same point about the architecture of mathematics. See the "Preface" to *Principia Mathematica*: ". . . the chief reason in favour of any theory on the principles of mathematics must always be inductive, i.e., it must lie in the fact that the theory in question enables us to deduce ordinary mathematics. In mathematics, the greatest degree of self-evidence is usually not to be found quite at the beginning, but at some later point; hence the early deductions, until they reach this point, give reasons rather for believing the premises because true consequences follow from them, than for believing the consequences because they follow from the premises." Contemporary preferences for deductive formalisms frequently blind us to this important fact, which is no less true today than it was in 1910.

explains why. Resemblance in behavior of systems without identity of the inner systems is particularly feasible if the aspects in which we are interested arise out of the *organization* of the parts, independently of all but a few properties of the individual components. Thus for many purposes we may be interested in only such characteristics of a material as its tensile and compressive strength. We may be profoundly unconcerned about its chemical properties, or even whether it is wood or iron.

The motor control patent cited earlier illustrates this abstraction to organizational properties. The invention consisted of a "combination" of "reversing means," of "field weakening means," that is to say, of components specified in terms of their functioning in the organized whole. How many ways are there of reversing a motor, or of weakening its field strength? We can simulate the system described in the patent claims in many ways without reproducing even approximately the actual physical device that is depicted. With a small additional step of abstraction, the patent claims could be restated to encompass mechanical as well as electrical devices. I suppose that any undergraduate engineer at Berkeley, Carnegie Mellon University, or MIT could design a mechanical system embodying reversibility and variable starting torque so as to simulate the system of the patent.

The Computer as Artifact

No artifact devised by man is so convenient for this kind of functional description as a digital computer. It is truly protean, for almost the only ones of its properties that are detectable in its behavior (when it is operating properly!) are the organizational properties. The speed with which it performs it basic operations may allow us to infer a little about its physical components and their natural laws; speed data, for example, would allow us to rule out certain kinds of "slow" components. For the rest, almost no interesting statement that one can make about an operating computer bears any particular relation to the specific nature of the hardware. A computer is an organization of elementary functional components in which, to a high approximation, only the function performed by those components is relevant to the behavior of the whole system.[14]

Computers as Abstract Objects

This highly abstractive quality of computers makes it easy to introduce mathematics into the study of their theory—and has led some to the erroneous conclusion that, as a computer science emerges, it will necessarily be a mathematical rather than an empirical science. Let me take up these two points in turn: the relevance of mathematics to computers and the possibility of studying computers empirically.

Some important theorizing, initiated by John von Neumann, has been done on the topic of computer reliability. The question is how to build a reliable system from unreliable parts. Notice that this is not posed as a question of physics or physical engineering. The components engineer is assumed to have done his best, but the parts are still unreliable! We can cope with the unreliability only by our manner of organizing them.

To turn this into a meaningful problem, we have to say a little more about the nature of the unreliable parts. Here we are aided by the knowledge that *any* computer can be assembled out of a small array of simple, basic elements. For instance, we may take as our primitives the so-called Pitts-McCulloch neurons. As their name implies, these components were devised in analogy to the supposed anatomical and functional characteristics of neurons in the brain, but they are highly abstracted. They are formally isomorphic with the simplest kinds of switching circuits—"and," "or," and "not" circuits. We postulate, now, that we are to build a system from such elements and that each elementary part has a specified probability of functioning correctly. The problem is to arrange the elements and their interconnections in such a way that the complete system will perform reliably.

The important point for our present discussion is that the parts could as well be neurons as relays, as well relays as transistors. The natural laws governing relays are very well known, while the natural laws governing

14. On the subject of this and the following paragraphs, see M. L. Minsky, *op. cit.*; then John von Neumann, "Probabilistic Logics and the Synthesis of Reliable Organisms from Unreliable Components," in C. E. Shannon and J. McCarthy (eds.), *Automata Studies* (Princeton: Princeton University Press, 1956).

neurons are known most imperfectly. But that does not matter, for all that is relevant for the theory is that the components have the specified level of unreliability and be interconnected in the specified way.

This example shows that the possibility of building a mathematical theory of a system or of simulating that system does not depend on having an adequate microtheory of the natural laws that govern the system components. Such a microtheory might indeed be simply irrelevant.

Computers as Empirical Objects

We turn next to the feasibility of an *empirical* science of computers—as distinct from the solid-state physics or physiology of their componentry.[15] As a matter of empirical fact almost all of the computers that have been designed have certain common organizational features. They almost all can be decomposed into an active processor (Babbage's "Mill") and a memory (Babbage's "Store") in combination with input and output devices. (Some of the larger systems, somewhat in the manner of colonial algae, are assemblages of smaller systems having some or all of these components. But perhaps I may oversimplify for the moment.) They are all capable of storing symbols (program) that can be interpreted by a program-control component and executed. Almost all have exceedingly limited capacity for simultaneous, parallel activity—they are basically one-thing-at-a-time systems. Symbols generally have to be moved from the larger memory components into the central processor before they can be acted upon. The systems are capable of only simple basic actions: recoding symbols, storing symbols, copying symbols, moving symbols, erasing symbols, and comparing symbols.

Since there are now many such devices in the world, and since the properties that describe them also appear to be shared by the human central nervous system, nothing prevents us from developing a natural history of them. We can study them as we would rabbits or chipmunks and discover how they behave under different patterns of environmental stimulation. Insofar as their behavior reflects largely the broad functional

15. A. Newell and H. A. Simon, "Computer Science as Empirical Inquiry," *Communications of the ACM, 19*(March 1976):113–126. See also H. A. Simon, "Artificial Intelligence: An Empirical Science," *Artificial Intelligence, 77*(1995): 95–127.

characteristics we have described, and is independent of details of their hardware, we can build a general—but empirical—theory of them.

The research that was done to design computer time-sharing systems is a good example of the study of computer behavior as an empirical phenomenon. Only fragments of theory were available to guide the design of a time-sharing system or to predict how a system of a specified design would actually behave in an environment of users who placed their several demands upon it. Most actual designs turned out initially to exhibit serious deficiencies, and most predictions of performance were startlingly inaccurate.

Under these circumstances the main route open to the development and improvement of time-sharing systems was to build them and see how they behaved. And this is what was done. They were built, modified, and improved in successive stages. Perhaps theory could have anticipated these experiments and made them unnecessary. In fact it didn't, and I don't know anyone intimately acquainted with these exceedingly complex systems who has very specific ideas as to how it might have done so. To understand them, the systems had to be constructed, and their behavior observed.[16]

In a similar vein computer programs designed to play games or to discover proofs for mathematical theorems spend their lives in exceedingly large and complex task environments. Even when the programs themselves are only moderately large and intricate (compared, say, with the monitor and operating systems of large computers), too little is known about their task environments to permit accurate prediction of how well they will perform, how selectively they will be able to search for problem solutions.

Here again theoretical analysis must be accompanied by large amounts of experimental work. A growing literature reporting these experiments is beginning to give us precise knowledge about the degree of heuristic power of particular heuristic devices in reducing the size of the problem spaces that must be searched. In theorem proving, for example, there has

16. The empirical, exploratory flavor of computer research is nicely captured by the account of Maurice V. Wilkes in his 1967 Turing Lecture, "Computers Then and Now," *Journal of the Association for Computing Machinery, 15*(January 1968):1–7.

been a whole series of advances in heuristic power based on and guided by empirical exploration: the use of the Herbrand theorem, the resolution principle, the set-of-support principle, and so on.[17]

Computers and Thought

As we succeed in broadening and deepening our knowledge—theoretical and empirical—about computers, we discover that in large part their behavior is governed by simple general laws, that what appeared as complexity in the computer program was to a considerable extent complexity of the environment to which the program was seeking to adapt its behavior.

This relation of program to environment opened up an exceedingly important role for computer simulation as a tool for achieving a deeper understanding of human behavior. For if it is the organization of components, and not their physical properties, that largely determines behavior, and if computers are organized somewhat in the image of man, then the computer becomes an obvious device for exploring the consequences of alternative organizational assumptions for human behavior. Psychology could move forward without awaiting the solutions by neurology of the problems of component design—however interesting and significant these components turn out to be.

Symbol Systems: Rational Artifacts

The computer is a member of an important family of artifacts called symbol systems, or more explicitly, physical symbol systems.[18] Another important member of the family (some of us think, anthropomorphically, it is the *most* important) is the human mind and brain. It is with this family

17. Note, for example, the empirical data in Lawrence Wos, George A. Robinson, Daniel F. Carson, and Leon Shalla, "The Concept of Demodulation in Theorem Proving," *Journal of the Association for Computing Machinery,* 14(October 1967):698–709, and in several of the earlier papers referenced there. See also the collection of programs in Edward Feigenbaum and Julian Feldman (eds.), *Computers and Thought* (New York: McGraw-Hill, 1963). It is common practice in the field to title papers about heuristic programs, "Experiments with an *XYZ* Program."

18. In the literature the phrase *information-processing system* is used more frequently than symbol system. I will use the two terms as synonyms.

of artifacts, and particularly the human version of it, that we will be primarily concerned in this book. Symbol systems are almost the quintessential artifacts, for adaptivity to an environment is their whole *raison d'être*. They are goal-seeking, information-processing systems, usually enlisted in the service of the larger systems in which they are incorporated.

Basic Capabilities of Symbol Systems

A physical symbol system holds a set of entities, called symbols. These are physical patterns (e.g., chalk marks on a blackboard) that can occur as components of symbol structures (sometimes called "expressions"). As I have already pointed out in the case of computers, a symbol system also possesses a number of simple processes that operate upon symbol structures—processes that create, modify, copy, and destroy symbols. A physical symbol system is a machine that, as it moves through time, produces an evolving collection of symbol structures.[19] Symbol structures can, and commonly do, serve as internal representations (e.g., "mental images") of the environments to which the symbol system is seeking to adapt. They allow it to model that environment with greater or less veridicality and in greater or less detail, and consequently to reason about it. Of course, for this capability to be of any use to the symbol system, it must have windows on the world and hands, too. It must have means for acquiring information from the external environment that can be encoded into internal symbols, as well as means for producing symbols that initiate action upon the environment. Thus it must use symbols to *designate* objects and relations and actions in the world external to the system.

Symbols may also designate processes that the symbol system can interpret and execute. Hence the programs that govern the behavior of a symbol system can be stored, along with other symbol structures, in the system's own memory, and executed when activated.

Symbol systems are called "physical" to remind the reader that they exist as real-world devices, fabricated of glass and metal (computers) or flesh and blood (brains). In the past we have been more accustomed to thinking of the symbol systems of mathematics and logic as abstract and disembodied, leaving out of account the paper and pencil and human minds that were required actually to bring them to life. Computers have

19. Newell and Simon, "Computer Science as Empirical Inquiry," p. 116.

transported symbol systems from the platonic heaven of ideas to the empirical world of actual processes carried out by machines or brains, or by the two of them working together.

Intelligence as Computation

The three chapters that follow rest squarely on the hypothesis that intelligence is the work of symbol systems. Stated a little more formally, the hypothesis is that a physical symbol system of the sort I have just described has the necessary and sufficient means for general intelligent action.

The hypothesis is clearly an empirical one, to be judged true or false on the basis of evidence. One task of chapters 3 and 4 will be to review some of the evidence, which is of two basic kinds. On the one hand, by constructing computer programs that are demonstrably capable of intelligent action, we provide evidence on the sufficiency side of the hypothesis. On the other hand, by collecting experimental data on human thinking that tend to show that the human brain operates as a symbol system, we add plausibility to the claims for necessity, for such data imply that all known intelligent systems (brains and computers) are symbol systems.

Economics: Abstract Rationality

As prelude to our consideration of human intelligence as the work of a physical symbol system, chapter 2 introduces a heroic abstraction and idealization—the idealization of human rationality which is enshrined in modern economic theories, particularly those called neoclassical. These theories are an idealization because they direct their attention primarily to the external environment of human thought, to decisions that are optimal for realizing the adaptive system's goals (maximization of utility or profit). They seek to define the decisions that would be substantively rational in the circumstances defined by the outer environment.

Economic theory's treatment of the limits of rationality imposed by the inner environment—by the characteristics of the physical symbol system—tends to be pragmatic, and sometimes even opportunistic. In the more formal treatments of general equilibrium and in the so-called "rational expectations" approach to adaptation, the possibilities that an information-processing system may have a very limited capability for adaptation are almost ignored. On the other hand, in discussions of the rationale for market mechanisms and in many theories of decision making under uncertainty, the procedural aspects of rationality receive more serious treatment.

In chapter 2 we will see examples both of neglect for and concern with the limits of rationality. From the idealizations of economics (and some criticisms of these idealizations) we will move, in chapters 3 and 4, to a more systematic study of the inner environment of thought—of thought processes as they actually occur within the constraints imposed by the parameters of a physical symbol system like the brain.

2

Economic Rationality: Adaptive Artifice

Because scarcity is a central fact of life—land, money, fuel, time, attention, and many other things are scarce—it is a task of rationality to allocate scarce things. Performing that task is the focal concern of economics.

Economics exhibits in purest form the artificial component in human behavior, in individual actors, business firms, markets, and the entire economy. The outer environment is defined by the behavior of other individuals, firms, markets, or economies. The inner environment is defined by an individual's, firm's, market's, or economy's goals and capabilities for rational, adaptive behavior. Economics illustrates well how outer and inner environment interact and, in particular, how an intelligent system's adjustment to its outer environment (its *substantive rationality*) is limited by its ability, through knowledge and computation, to discover appropriate adaptive behavior (its *procedural rationality*).

The Economic Actor

In the textbook theory of the business firm, an "entrepreneur" aims at maximizing profit, and in such simple circumstances that the computational ability to find the maximum is not in question. A cost curve relates dollar expenditures to amount of product manufactured, and a revenue curve relates income to amount of product sold. The goal (maximizing the difference between income and expenditure) fully defines the firm's inner environment. The cost and revenue curves define the outer environment.[1] Elementary calculus shows how to find the profit-maximizing

---

1. I am drawing the line between outer and inner environment not at the firm's boundary but at the skin of the entrepreneur, so that the factory is part of the external technology; the brain, perhaps assisted by computers, is the internal.

quantity by taking a derivative (rate at which profit changes with change in quantity) and setting it equal to zero.

Here are all the elements of an artificial system adapting to an outer environment, subject only to the goal defined by the inner environment. In contrast to a situation where the adaptation process is itself problematic, we can predict the system's behavior without knowing how it actually computes the optimal output. We need consider only substantive rationality.[2]

We can interpret this bare-bones theory of the firm either positively (as describing how business firms behave) or normatively (as advising them how to maximize profits). It is widely taught in both senses in business schools and universities, just as if it described what goes on, or could go on, in the real world. Alas, the picture is far too simple to fit reality.

Procedural Rationality

The question of maximizing the difference between revenue and cost becomes interesting when, in more realistic circumstances, we ask how the firm actually goes about discovering that maximizing quantity. Cost accounting may estimate the approximate cost of producing any particular output, but how much can be sold at a specific price and how this amount varies with price (the elasticity of demand) usually can be guessed only roughly. When there is uncertainty (as there always is), prospects of profit must be balanced against risk, thereby changing profit maximization to the much more shadowy goal of maximizing a profit-vs.-risk "utility function" that is assumed to lurk somewhere in the recesses of the entrepreneur's mind.

But in real life the business firm must also choose product quality and the assortment of products it will manufacture. It often has to invent and design some of these products. It must schedule the factory to produce a profitable combination of them and devise marketing procedures and structures to sell them. So we proceed step by step from the simple caricature of the firm depicted in the textbooks to the complexities of real firms in the real world of business. At each step toward realism, the problem

2. H. A. Simon, "Rationality as Process and as Product of Thought," *American Economic Review*, 68(1978):1–16.

gradually changes from choosing the right course of action (substantive rationality) to finding a way of calculating, very approximately, where a good course of action lies (procedural rationality). With this shift, the theory of the firm becomes a theory of estimation under uncertainty and a theory of computation—decidedly non-trivial theories as the obscurities and complexities of information and computation increase.

Operations Research and Management Science

Today several branches of applied science assist the firm to achieve procedural rationality.[3] One of them is operations research (OR); another is artificial intelligence (AI). OR provides algorithms for handling difficult multivariate decision problems, sometimes involving uncertainty. Linear programming, integer programming, queuing theory, and linear decision rules are examples of widely used OR procedures.

To permit computers to find optimal solutions with reasonable expenditures of effort when there are hundreds or thousands of variables, the powerful algorithms associated with OR impose a strong mathematical structure on the decision problem. Their power is bought at the cost of shaping and squeezing the real-world problem to fit their computational requirements: for example, replacing the real-world criterion function and constraints with linear approximations so that linear programming can be used. Of course the decision that is optimal for the simplified approximation will rarely be optimal in the real world, but experience shows that it will often be satisfactory.

The alternative methods provided by AI, most often in the form of heuristic search (selective search using rules of thumb), find decisions that are "good enough," that *satisfice*. The AI models, like OR models, also only approximate the real world, but usually with much more accuracy and detail than the OR models can admit. They can do this because heuristic search can be carried out in a more complex and less well structured problem space than is required by OR maximizing tools. The price paid

3. For a brief survey of these developments, see H. A. Simon, "On How to Decide What to Do," *The Bell Journal of Economics,* 9(1978):494–507. For an estimate of their impact on management, see H. A. Simon, *The New Science of Management Decision,* rev. ed. (Englewood Cliffs, NJ: Prentice-Hall, 1977), chapters 2 and 4.

for working with the more realistic but less regular models is that AI methods generally find only satisfactory solutions, not optima. We must trade off satisficing in a nearly-realistic model (AI) against optimizing in a greatly simplified model (OR). Sometimes one will be preferred, sometimes the other.

AI methods can handle combinatorial problems (e.g., factory scheduling problems) that are beyond the capacities of OR methods, even with the largest computers. Heuristic methods provide an especially powerful problem-solving and decision-making tool for humans who are unassisted by any computer other than their own minds, hence must make radical simplifications to find even approximate solutions. AI methods also are not limited, as most OR methods are, to situations that can be expressed quantitatively. They extend to all situations that can be represented symbolically, that is, verbally, mathematically or diagrammatically.

OR and AI have been applied mainly to business decisions at the middle levels of management. A vast range of top management decisions (e.g., strategic decisions about investment, R&D, specialization and diversification, recruitment, development, and retention of managerial talent) are still mostly handled traditionally, that is, by experienced executives' exercise of judgment.

As we shall see in chapters 3 and 4, so-called "judgment" turns out to be mainly a non-numerical heuristic search that draws upon information stored in large expert memories. Today we have learned how to employ AI techniques in the form of so-called *expert systems* in a growing range of domains previously reserved for human expertise and judgment—for example, medical diagnosis and credit evaluation. Moreover, while classical OR tools could only choose among predefined alternatives, AI expert systems are now being extended to the generation of alternatives, that is, to problems of design. More will be said about these developments in chapters 5 and 6.

Satisficing and Aspiration Levels

What a person *cannot* do he or she *will not* do, no matter how strong the urge to do it. In the face of real-world complexity, the business firm turns to procedures that find good enough answers to questions whose best answers are unknowable. Because real-world optimization, with or without computers, is impossible, the real economic actor is in fact a satisficer, a person who accepts "good enough" alternatives, not because less is preferred to more but because there is no choice.

Many economists, Milton Friedman being perhaps the most vocal, have argued that the gap between satisfactory and best is of no great importance, hence the unrealism of the assumption that the actors optimize does not matter; others, including myself, believe that it does matter, and matters a great deal.[4] But reviewing this old argument would take me away from my main theme, which is to show how the behavior of an artificial system may be strongly influenced by the limits of its adaptive capacities—its knowledge and computational powers.

One requirement of optimization not shared by satisficing is that all alternatives must be measurable in terms of a common utility function. A large body of evidence shows that human choices are not consistent and transitive, as they would be if a utility function existed.[5] But even in a satisficing theory we need some criteria of satisfaction. What realistic measures of human profit, pleasure, happiness and satisfaction can serve in place of the discredited utility function?

Research findings on the psychology of choice, indicate some properties a thermometer of satisfaction should have. First, unlike the utility function, it is not limited to positive values, but has a zero point (of minimal contentment). Above zero, various degrees of satisfaction are experienced, and below zero, various degrees of dissatisfaction. Second, if periodic readings are taken of people in relatively stable life circumstances, we only occasionally find temperatures very far from zero in either direction, and the divergent measurements tend to regress over time back toward the zero mark. Most people consistently register either slightly below zero (mild discontent) or a little above (moderate satisfaction).

4. I have argued the case in numerous papers. Two recent examples are "Rationality in Psychology and Economics," *The Journal of Business,* 59(1986):S209-S224 (No. 4, Pt. 2); and "The State of Economic Science," in W. Sichel (ed.), *The State of Economic Science* (Kalamazoo, MI: W. E. Upjohn Institute for Employment Research, 1989).

5. See, for example, D. Kahneman and A. Tversky, "On the Psychology of Prediction," *Psychological Review,* 80(1973):237–251, and H. Kunreuther et al., *Disaster Insurance Protection* (New York: Wiley, 1978).

To deal with these phenomena, psychology employs the concept of *aspiration level*. Aspirations have many dimensions: one can have aspirations for pleasant work, love, good food, travel, and many other things. For each dimension, expectations of the attainable define an aspiration level that is compared with the current level of achievement. If achievements exceed aspirations, satisfaction is recorded as positive; if aspirations exceed achievements, there is dissatisfaction. There is no simple mechanism for comparison *between* dimensions. In general a large gain along one dimension is required to compensate for a small loss along another—hence the system's net satisfactions are history-dependent, and it is difficult for people to balance compensatory offsets.

Aspiration levels provide a computational mechanism for satisficing. An alternative satisfices if it meets aspirations along all dimensions. If no such alternative is found, search is undertaken for new alternatives. Meanwhile, aspirations along one or more dimensions drift down gradually until a satisfactory new alternative is found or some existing alternative satisfices. A theory of choice employing these mechanisms acknowledges the limits on human computation and fits our empirical observations of human decision making far better than the utility maximization theory.[6]

Markets and Organizations

Economics has been concerned less with individual consumers or business firms than with larger artificial systems: the economy and its major components, markets. Markets aim to coordinate the decisions and behavior of multitudes of economic actors—to guarantee that the quantity of brussels sprouts shipped to market bears some reasonable relation to the quantity that consumers will buy and eat, and that the price at which brussels sprouts can be sold bears a reasonable relation to the cost of producing them. Any society that is not a subsistence economy, but has

6. H. A. Simon, "A Behavioral Model of Rational Choice," *Quarterly Journal of Economics*, 6(1955):99–118; I. N. Gallhofer and W. E. Saris, *Foreign Policy Decision-Making: A Qualitative and Quantitative Analysis of Political Argumentation* (New York: Praeger, in press).

substantial specialization and division of labor, needs mechanisms to perform this coordinative function.

Markets are only one, however, among the spectrum of mechanisms of coordination on which any society relies. For some purposes, central planning based on statistics provides the basis for coordinating behavior patterns. Highway planning, for example, relies on estimates of road usage that reflect statistically stable patterns of driving behavior. For other purposes, bargaining and negotiation may be used to coordinate individual behaviors, for instance, to secure wage agreements between employers and unions or to form legislative majorities. For still other coordinative functions, societies employ hierarchic organizations—business, governmental and educational—with lines of formal authority running from top to bottom and networks of communications lacing through the structure. Finally, for making certain important decisions and for selecting persons to occupy positions of public authority, societies employ a wide variety of balloting procedures.

Although all of these coordinating techniques can be found somewhere in almost any society, their mix and applications vary tremendously from one nation or culture to another.[7] We ordinarily describe capitalist societies as depending mostly on markets for coordination and socialist societies as depending mostly on hierarchic organizations and planning, but this is a gross oversimplification, for it ignores the uses of voting in democratic societies of either kind, and it ignores the great importance of large organizations in modern "market" societies.

The economic units in capitalist societies are mostly business firms, which are themselves hierarchic organizations, some of enormous size, that make almost negligible use of markets in their internal functioning. Roughly eighty percent of the human economic activity in the American economy, usually regarded as almost the epitome of a "market" economy, takes place in the internal environments of business and other organizations and not in the external, between-organization environments of markets.[8] To avoid misunderstanding, it would be appropriate to call such

7. R. A. Dahl and C. E. Lindblom, *Politics, Economics, and Welfare* (New York: Harper and Brothers, 1953).

8. H. A. Simon, "Organizations and Markets," *Journal of Economic Perspectives*, 5(1991):25–44.

a society an organization-&-market economy; for in order to give an account of it we have to pay as much attention to organizations as to markets.

The Invisible Hand

In examining the processes of social coordination, economics has given top billing—sometimes almost exclusive billing—to the market mechanism. It is indeed a remarkable mechanism which under many circumstances can bring it about that the producing, consuming, buying and selling behaviors of enormous numbers of people, each responding only to personal selfish interests, allocate resources so as to clear markets—do in fact nearly balance the production with the consumption of brussels sprouts and all the other commodities the economy produces and uses.

Only relatively weak conditions need be satisfied to bring about such an equilibrium. Achieving it mainly requires that prices drop in the face of an excess supply, and that quantities produced decline when prices are lowered or when inventories mount. Any number of dynamic systems can be formulated that have these properties, and these systems will seek equilibrium and oscillate stably around it over a wide range of conditions.

There have been many recent laboratory experiments on market behavior, sometimes with human subjects, sometimes with computer programs as simulated subjects.[9] Experimental markets in which the simulated traders are "stupid" sellers, knowing only a minimum price below which they should not sell, and "stupid" buyers, knowing only a maximum price above which they should not buy move toward equilibrium almost as rapidly as markets whose agents are rational in the classical sense.[10]

Markets and Optimality.   These findings undermine the much stronger claims that are made for the price mechanism by contemporary neoclassical economics. Claims that it does more than merely clear markets require the strong assumptions of perfect competition and of maximization of

9. V. L. Smith, *Papers in Experimental Economics* (New York: Cambridge University Press, 1991.)

10. D. J. Gode and S. Sunder, "Allocative Efficiency of Markets with Zero Intelligence Traders," *Journal of Political Economy,* 101(1993):119–127.

profit or utility by the economic actors. With these assumptions, but not without them, the market equilibrium can be shown to be optimal in the sense that it could not be altered so as to make everyone simultaneously better off. These are the familiar propositions of Pareto optimality of competitive equilibrium that have been formalized so elegantly by Arrow, Debreu, Hurwicz, and others.[11]

The optimality theorems stretch credibility, so far as real-world markets are concerned, because they require substantive rationality of the kinds we found implausible in our examination of the theory of the firm. Markets populated by consumers and producers who satisfice instead of optimizing do not meet the conditions on which the theorems rest. But the experimental data on simulated markets show that market clearing, the only property of markets for which there is solid empirical evidence, can be achieved without the optimizing assumptions, hence also without claiming that markets do produce a Pareto optimum. As Samuel Johnson said of the dancing dog, "The marvel is not that it dances well, but that it dances at all"—the marvel is not that markets optimize (they don't) but that they often clear.

Order without a Planner. We have become accustomed to the idea that a natural system like the human body or an ecosystem regulates itself. This is in fact a favorite theme of the current discussion of complexity which we will take up in later chapters. We explain the regulation by feedback loops rather than a central planning and directing body. But somehow, untutored intuitions about self-regulation without central direction do not carry over to the artificial systems of human society. I retain vivid memories of the astonishment and disbelief expressed by the architecture students to whom I taught urban land economics many years ago when I pointed to medieval cities as marvelously patterned systems that had mostly just "grown" in response to myriads of individual human decisions. To my students a pattern implied a planner in whose mind it had been conceived and by whose hand it had been implemented. The idea that a city could acquire its pattern as naturally as a snowflake was

11. See Gerard Debreu, *Theory of Value: An Axiomatic Analysis of Economic Equilibrium* (New York: Wiley, 1959).

foreign to them. They reacted to it as many Christian fundamentalists responded to Darwin: no design without a Designer!

Marxist fundamentalists reacted in a similar way when, after World War I, they undertook to construct the new socialist economies of eastern Europe. It took them some thirty years to realize that markets and prices might play a constructive role in socialist economies and might even have important advantages over central planning as tools for the allocation of resources. My sometime teacher, Oscar Lange, was one of the pioneers who carried this heretical notion to Poland after the Second World War and risked his career and his life for the idea.

With the collapse of the Eastern European economies around 1990 the simple faith in central planning was replaced in some influential minds by an equally simple faith in markets. The collapse taught that modern economies cannot function well without smoothly operating markets. The poor performance of these economies since the collapse has taught that they also cannot function well without effective organizations.

If we focus on the equilibrating functions of markets and put aside the illusions of Pareto optimality, market processes commend themselves primarily because they avoid placing on a central planning mechanism a burden of calculation that such a mechanism, however well buttressed by the largest computers, could not sustain. Markets appear to conserve information and calculation by assigning decisions to actors who can make them on the basis of information that is available to them locally— that is, without knowing much about the rest of the economy apart from the prices and properties of the goods they are purchasing and the costs of the goods they are producing.

No one has characterized market mechanisms better than Friederich von Hayek who, in the decades after World War II, was their leading interpreter and defender. His defense did not rest primarily upon the supposed optimum attained by them but rather upon the limits of the inner environment—the computational limits of human beings:[12]

The most significant fact about this system is the economy of knowledge with which it operates, or how little the individual participants need to know in order to be able to take the right action.

12. F. von Hayek, "The Use of Knowledge in Society," *American Economic Review*, 35(September 1945):519–30, at p. 520.

The experiments on simulated markets, described earlier, confirm his view. At least under some circumstances, market traders using a very small amount of mostly local information and extremely simple (and non-optimizing) decision rules, can balance supply and demand and clear markets.

It is time now that we turn to the role of organizations in an organization-&-market economy and the reasons why all economic activities are not left to market forces. In preparation for this topic, we need to look at the phenomena of uncertainty and expectations.

Uncertainty and Expectations

Because the consequences of many actions extend well into the future, correct prediction is essential for objectively rational choice. We need to know about changes in the natural environment: the weather that will affect next year's harvest. We need to know about changes in social and political environments beyond the economic: the civil warfare of Bosnia or Sri Lanka. We need to know about the future behaviors of other economic actors—customers, competitors, suppliers—which may be influenced in turn by our own behaviors.

In simple cases uncertainty arising from exogenous events can be handled by estimating the probabilities of these events, as insurance companies do—but usually at a cost in computational complexity and information gathering. An alternative is to use feedback to correct for unexpected or incorrectly predicted events. Even if events are imperfectly anticipated and the response to them less than accurate, adaptive systems may remain stable in the face of severe jolts, their feedback controls bringing them back on course after each shock that displaces them. After we fail to predict the blizzard, snowplows still clear the streets. Although the presence of uncertainty does not make intelligent choice impossible, it places a premium on robust adaptive procedures instead of optimizing strategies that work well only when finely tuned to precisely known environments.[13]

13. A remarkable paper by Kenneth Arrow, reprinted in *The New Palgrave: A Dictionary of Economics* (London: Macmillan Press, 1987), v. 2, pp. 69–74, under the title of "Economic Theory and the Hypothesis of Rationality," shows that to preserve the Pareto optimality properties of markets when there is uncertainty

Expectations. A system can generally be steered more accurately if it uses feedforward, based on prediction of the future, in combination with feedback, to correct the errors of the past. However, forming expectations to deal with uncertainty creates its own problems. Feedforward can have unfortunate destabilizing effects, for a system can overreact to its predictions and go into unstable oscillations. Feedforward in markets can become especially destabilizing when each actor tries to anticipate the actions of the others (and hence their expectations).

The standard economic example of destabilizing expectations is the speculative bubble. Bubbles that ultimately burst are observed periodically in the world's markets (the Tulip Craze being one of many well-known historical examples). Moreover, bubbles and their bursts have now been observed in experimental markets, the overbidding occurring even though subjects know that the market must again fall to a certain level on a specified and not too distant date.

Of course not all speculation blows bubbles. Under many circumstances market speculation stabilizes the system, causing its fluctuations to become smaller, for the speculator attempts to notice when particular prices are above or below their "normal" or equilibrium levels in order to sell or buy, respectively. Such actions push the prices closer to equilibrium.

Sometimes, however, a rising price creates the expectation that it will go higher yet, hence induces buying rather than selling. There ensues a game of economic "chicken," all the players assuming that they can get out just before the crash occurs. There is general consensus in economics that destabilizing expectations play an important role in monetary hyperinflation and in the business cycle. There is less consensus as to *whose* expectations are the first movers in the chain of reactions or what to do about it.

The difficulties raised by mutual expectations appear wherever markets are not perfectly competitive. In perfect competition, each firm assumes that market prices cannot be affected by their actions: prices are as much a part of the external environment as are the laws of the physical world.

---

about the future, we must impose information and computational requirements on economic actors that are exceedingly burdensome and unrealistic.

But in the world of imperfectly competitive markets, firms need not make this assumption. If, for example, there are only a few firms in an industry, each may try to outguess its competitors. If more than one plays this game, even the definition of rationality comes into question.

The Theory of Games.   A century and a half ago, Augustin Cournot undertook to construct a theory of rational choice in markets involving two firms.[14] He assumed that each firm, with *limited* cleverness, formed an expectation of its competitor's reaction to its actions, but that each carried the analysis only one move deep. But what if one of the firms, or both, tries to take into account the reactions to the reactions? They may be led into an infinite regress of outguessing.

A major step toward a clearer formulation of the problem was taken a century later, in 1944, when von Neumann and Morgenstern published *The Theory of Games and Economic Behavior.*[15] But far from solving the problem, the theory of games demonstrated how intractable a task it is to prescribe optimally rational action in a multiperson situation where interests are opposed.

The difficulty of defining rationality exhibits itself well in the so-called Prisoners' Dilemma game.[16] In the Prisoners' Dilemma, each player has a choice between two moves, one cooperative and one aggressive. If both choose the cooperative move, both receive a moderate reward. If one chooses the cooperative move, but the other the aggressive move, the co-operator is penalized severely while the aggressor receives a larger reward. If both choose the aggressive move, both receive lesser penalties. There is no obvious rational strategy. Each player will gain from cooperation if and only if the partner does not aggress, but each will gain even more from aggression if he can count on the partner to cooperate. Treachery pays, unless it is met with treachery. The mutually beneficial strategy is unstable.

14. *Researches into the Mathematical Principles of the Theory of Wealth* (New York: Augustus M. Kelley, 1960), first published in 1838.

15. Princeton: Princeton University Press, 1944.

16. R. D. Luce and H. Raiffa, *Games and Decisions* (New York: Wiley, 1957), pp. 94–102; R. M. Axelrod, *The Evolution of Cooperation,* (New York: Basic Books, 1984).

Are matters improved by playing the game repetitively? Even in this case, cleverly timed treachery pays off, inducing instability in attempts at cooperation. However, in actual experiments with the game, it turns out that cooperative behavior occurs quite frequently, and that a tit-for-tat strategy (behave cooperatively until the other player aggresses; then aggress once but return to cooperation if the other player also does) almost always yields higher rewards than other strategies. Roy Radner has shown (personal communication) that if players are striving for a *satisfactory* payoff rather than an *optimal* payoff, the cooperative solution can be stable. Bounded rationality appears to produce better outcomes than unbounded rationality in this kind of competitive situation.

The Prisoners' Dilemma game, which has obvious real-world analogies in both politics and business, is only one of an unlimited number of games that illustrates the paradoxes of rationality wherever the goals of the different actors conflict totally or partially. Classical economics avoided these paradoxes by focusing upon the two situations (monopoly and perfect competition) where mutual expectations play no role.

Market institutions are workable (but not optimal) well beyond that range of situations precisely because the limits on human abilities to compute possible scenarios of complex interaction prevent an infinite regress of mutual outguessing. Game theory's most valuable contribution has been to show that rationality is effectively undefinable when competitive actors have unlimited computational capabilities for outguessing each other, but that the problem does not arise as acutely in a world, like the real world, of bounded rationality.

Rational Expectations.   A different view from the one just expressed was for a time popular in economics: that the problem of mutual outguessing should be solved by assuming that economic actors form their expectations "rationally."[17] This is interpreted to mean that the actors know (and agree on) the laws that govern the economic system and that their predic-

17. The idea and the phrase "rational expectations" originated with J. F. Muth, "Rational Expectations and the Theory of Price Movements," *Econometrica*, 29(1961):315–335. The notion was picked up, developed, and applied systematically to macroeconomics by R. E. Lucas, Jr., E. C. Prescott, T. J. Sargent, and others.

tions of the future are unbiased estimates of the equilibrium defined by these laws. These assumptions rule out most possibilities that speculation will be destabilizing.

Although the assumptions underlying rational expectations are empirical assumptions, almost no empirical evidence supports them, nor is it obvious in what sense they are "rational" (i.e., utility maximizing). Business firms, investors, or consumers do not possess even a fraction of the knowledge or the computational ability required for carrying out the rational expectations strategy. To do so, they would have to share a model of the economy and be able to compute its equilibrium.

Today, most rational expectationists are retreating to more realistic schemes of "adaptive expectations," in which actors gradually learn about their environments from the unfolding of events around them.[18] But most approaches to adaptive expectations give up the idea of outguessing the market, and instead assume that the environment is a slowly changing "given" whose path will not be significantly affected by the decisions of any one actor.

In sum, our present understanding of the dynamics of real economic systems is grossly deficient. We are especially lacking in empirical information about how economic actors, with their bounded rationality, form expectations about the future and how they use such expectations in planning their own behavior. Economics could do worse than to return to the empirical methods proposed (and practiced) by George Katona for studying expectation formation,[19] and to an important extent, the current interest in experimental economics represents such a return. In face of the current gaps in our empirical knowledge there is little empirical basis for choosing among the competing models currently proposed by economics to account for business cycles, and consequently, little rational basis for choosing among the competing policy recommendations that flow from those models.

18. T. J. Sargent, *Bounded Rationality in Macroeconomics* (Oxford: Clarendon Press, 1993). Note that Sargent even borrows the label of "bounded rationality" for his version of adaptive expectations, but, regrettably, does not borrow the empirical methods of direct observation and experimentation that would have to accompany it in order to validate the particular behavioral assumptions he makes.

19. G. Katona, *Psychological Analysis of Economic Behavior* (New York: McGraw-Hill, 1951).

Business Organizations

We turn now to the great mass of economic activity that takes place within the internal environments of organizations. The key question here, one much discussed in "the new institutional economics" (NIE),[20] is: what determines the boundary between organizations and markets; when will one be used, and when the other, to organize economic activity?

The Organization-Market Boundary.   At the outset it should be observed that the boundary is often quite movable. For example, retail sales of automobiles are usually handled by dealerships, organizations with separate ownership from the manufacturers. Many other commodities are sold directly to consumers by manufacturers, and in some industries (e.g., fast foods) there is a combination of direct outlets and franchise agencies. The franchise is an excellent example of a hybrid species, as is the sole-source vendor who supplies raw materials or parts to a manufacturer.

We take the frequent movability or indefiniteness of organizational boundaries as evidence that often there is nearly a balance between the advantages of markets and organizations. Nevertheless we recall again the vast activity that takes place inside organizations, many of them very large, as an indication that in many circumstances they offer important advantages over markets.

The NIE explanation for sometimes preferring organizations to markets is that certain kinds of market contracts incur transaction costs that can be avoided or reduced by replacing the sales contract by an employment relation. On the other hand, as all economic actors are supposed by the NIE theory to be motivated by selfish interest, organizations incur the costs of rewarding their employees for following organizational goals instead of personal interest and of supervising them to see that they do so.[21]

This account of the relative advantages of the two institutions misses essential parts of the story, especially the opportunities for decentralization of decision making within organizations. These opportunities de-

20. O. E. Williamson, *Markets and Hierarchies* (New York: The Free Press, 1975).

21. O. E. Williamson, *op. cit.*; O. E. Williamson, *The Economic Institutions of Capitalism* (New York: The Free Press, 1985).

pend, in turn, upon the strength of the loyalties of employees to their organizations, and their identification with organizational objectives that derives from loyalty and from the local informational environment in which they find themselves.

Decentralization.   Organizations are not highly centralized structures in which all the important decisions are made at the center. Organizations operating in that centralized way would exceed the limits of human procedural rationality and lose many of the advantages attainable from the use of hierarchical authority. Real-world organizations behave quite differently.[22]

As a single decision may be influenced by a large number of facts and criteria of choice, some fraction of these premises may be specified by superiors without implying complete centralization. Organizations can localize and minimize information demands just as markets do, by decentralizing decisions. Matters of fact can be determined wherever the most skill and information is located to determine them, and they can then be communicated to "collecting points" where all the facts relevant to an issue can be put together and a decision reached. We can think of a decision as produced by executing a large computer program, each subroutine having its special tasks and relying on local sources of information. No single person or group need be expert on all aspects of the decision.

Thus business organizations, like markets, are vast distributed computers whose decision processes are substantially decentralized. The top level of a large corporation, which is typically subdivided into specialized product groups, will perform only a few functions, most often: (1) the "investment banking" function of allocating funds for capital projects, (2) selection of top executive personnel, and (3) long-range planning for capital funds and for possible new activities outside the scope of existing divisions.

Markets and organizations, however decentralized, are not fully equivalent in their effects. None of the theorems of optimality in resource

22. J. G. March and H. A. Simon, *Organizations,* 2nd ed. (Cambridge, MA: Blackwell, 1993).

allocation that are provable for ideal competitive markets can be proved for hierarchies, but this does not mean that real organizations operate inefficiently as compared to real markets.

Externalities.   Economists sometimes state the case for organizations as opposed to markets in terms of *externalities*. Externalities arise because the price mechanism works as advertised only when all of the inputs and outputs of an activity are subject to market pricing. A traditional example of an externality is a factory that is allowed to spew smoke from its stacks without compensating the surrounding homeowners. In these circumstances the price mechanism will not secure a socially desirable level of manufacturing activity; the product, priced below its social cost, will be overused.

The economist's preferred remedy for externalities is to bring the undesired consequences within the calculus of the price system: tax the emission of smoke, for example. This raises the question of how the tax is to be set. Although the techniques of cost-benefit analysis can provide answers, they are administrative answers and not answers given by an automatic market mechanism.

Similar questions of externalities among corporate divisional operations make large corporations less than fully willing to allow transactions among their component divisions and departments to be governed wholly by internal markets. In the absence of perfect competition, internal market prices are administered or negotiated prices, not competitive prices.

Uncertainty.   Uncertainty often persuades social systems to use hierarchy rather than markets in making decisions. It is not reasonable to allow the production department and the marketing department in the widget company to make independent estimates of next year's demand for widgets if the production department is to make the widgets that the marketing department is to sell. In matters like this, and also in matters of product design, it may be preferable that all the relevant departments operate on the *same* body of assumptions even if (or perhaps "especially if") the uncertainties might justify quite a range of different assumptions. In facing uncertainty, standardization and coordination, achieved through agreed-upon assumptions and specifications, may be more effective than prediction.

Uncertainty calls for flexibility, but markets do not always provide the greatest flexibility in the face of uncertainty. All depends on the sources of the uncertainty. If what is uncertain is a multitude of facts about individual and separate markets, then decentralized pricing will appear attractive; if the uncertainty encompasses major events that will affect many parts of the organization in the same direction, then it may be advantageous to centralize the making of assumptions about the future and to require the decentralized units to use these assumptions in their decisions.

Uncertainty is especially troublesome when it involves expectations by one unit about what other units in the same organization will do. Left to the market, this kind of uncertainty leads directly to the dilemmas of rationality that we described earlier in terms of game theory and rational expectations. Absorption of the uncertainty by the organization through managerial coordination may be the most effective course. We see in uncertainty a frequent source of advantage of organizations over markets as decision-making mechanisms.

In a world of bounded rationality there are several ways to magnify the computing capabilities of individual human beings and enhance the possibilities of their collective survival and prosperity. With the combined use of markets and administrative hierarchies, the human species has enormously increased its capabilities for specialization and division of work. It would be too much to attribute the vast growth and spread of human populations to such mechanisms alone—modern medicine and modern technology have had something to do with it too—but the (perhaps temporary) dominance of our species over the globe today is witness to the augmentation of human reason—applied to local, not global, concerns—that has been made possible by these social artifacts.

Organizational Loyalties and Identifications

Brief mention was made earlier of a crucial reason why so much human activity takes place within organizations: people acquire loyalty, and often a large amount of loyalty, to the groups, including organizations, to which they belong.

Consequences of Identification.   Organizational loyalty is perhaps better labeled *identification,* for it is both motivational and cognitive. The motivational component is an attachment to group goals and a willingness to work for them even at some sacrifice of personal goals. (In effect, the group goals *become* personal goals.) The ethnic conflict we observe in many parts of the world provides vivid evidence of this attachment to group goals and the differential treatment it generates between "we" and "they."

Identification with an organization also has a cognitive component, for members are surrounded by information, conceptions and frames of reference quite different from those of people outside the organization or in a different organization. As creatures of bounded rationality, incapable of dealing with the world in all of its complexity, we form a simplified picture of the world, viewing it from our particular organizational vantage point and our organization's interests and goals.

This frame of reference and information provided by an organization influence strongly the processing and outcomes of decisions. The frame of reference varies, too, from one organization unit to another and from one level to another, so that an employee may identify at one time with his department, at another with his section, at another with the whole company.

Affected by their organizational identifications, members frequently pursue organizational goals at the expense of their own interests—that is to say, behave in a way that is altruistic from a personal standpoint. No organization could survive that elicited only behavior for which employees felt selfishly rewarded and that supervisors could enforce. The added effort that is elicited by identification is a major and essential source of organizational effectiveness and is a principal reason for carrying out economic activities in organizations rather than markets.

Evolutionary Basis for Identification. It may be objected that human beings are basically selfish and do not behave in this altruistic fashion. In fact, neo-Darwinian evolutionary theory has generally claimed that altruism, except to close relatives, is inconsistent with the basic postulate that organisms evolve to increase their fitness.[23] However, I should like to show that this widely repeated claim is mistaken.[24]

23. The case is stated, for example, in R. Dawkins, *The Selfish Gene* (New York: Oxford University Press, 1989).

24. H. A. Simon, "A Mechanism for Social Selection and Successful Altruism," *Science*, 250(1990):1665–1668.

Because of their bounded rationality, and because they can therefore greatly enhance their limited knowledge and skill by accepting information and advice from the social groups to which they belong, individuals who are *docile*—who tend to accept such information and advice—have a great advantage in fitness over those who are not docile—who reject social influence. Docile people do not have to learn about hot stoves by touching them.

Most social influence does enhance the fitness of the recipient. It provides information and advice about the world that is generally valid—or at least much more informative and valid than the information the recipient could generate independently. But docility can be "taxed" by influencing people also to take certain actions that are not personally beneficial but are beneficial to the group. As long as the "taxation" is not so heavy as to cancel the advantages of docility, the altruistic individual will be fitter than the non-docile individual. By this means, the fitness of the organization will be enhanced by the docility, hence altruism, of its members. Although, docility is *generally* rewarding to the individual, some fraction of the behavior it induces is altruistic in this sense, and this altruism is an important factor in the efficacy of organizations.

We can summarize our account of the respective roles of markets and organizations in a modern society as follows: (1) organizations find their niches wherever constellations of interdependent activities are best carried out in coordinated fashion in order to remove the need for individuals' outguessing each other; (2) the human motivation that makes organizations viable and alleviates the public goods problems that arise when individual efforts cannot be tied closely to individual rewards is provided by organizational loyalty and identification; (3) in both organizations and markets, the bounds on human rationality are addressed by arranging decisions so that the steps in decision making can depend largely on information that is locally available to individuals.

The Evolutionary Model

Evolutionary processes are significant not only for explaining organizational loyalty, but also for describing and explaining the historical development of economic institutions, including business firms. The simplest scheme of evolution depends on two processes: a generator and a test. The generator produces variety, new forms that have not existed previously, whereas the test culls out the generated forms so that only those that are well fitted to the environment will survive. In modern biological Darwinism, genetic mutation and crossover of chromosomes are the principal generators, and natural selection is the test.

The Alternative Theory of Economic Man

No one supposes that a modern organization-&-market economy is the product of deliberate design. Surely it evolved from earlier subsistence economies, shaped by myriads of decisions made by hosts of actors over thousands of years. By contrast, most accounts of business firms assume that actors deliberately select actions appropriate to their goals within the context of the given economic environment. Adaptation, in the latter accounts, stems from selection by rational actors, not by natural selection of those actors whose behavior happens to be adaptive. An evolutionary theory of the firm might argue that it does not matter whether people maximize or satisfice, for in a world of competitive markets only those who make decisions *as if* they were maximizing will survive.[25] Does this evolutionary argument in fact imply optimization?

Our discussion will have implications for biology as well as economics, for evolutionary biology uses the language of optimality quite freely and in recent years has even borrowed linear programming and other OR techniques to predict the outcomes of natural selection in biological systems. This is legitimate only if optimization would lead reliably to the same equilibria as would natural selection.

Local and Global Maxima

For the question before us, the difference between local and global maxima is crucial. In the landscape of California every tiny hill is a local maxi-

25. A. A. Alchian, "Uncertainty, Evolution, and Economic Theory," *Journal of Political Economy,* 58(1950):211–222; M. Friedman, "The Methodology of Positive Economics," in *Essays in Positive Economics* (Chicago: University of Chicago Press, 1953). The identification of selection with optimization is challenged by S. G. Winter, for example, in his "Economic Natural Selection and the Theory of the Firm," *Yale Economic Essays,* 4(1964):225–272.

mum of altitude, but only Mt. Whitney is a global maximum. For many purposes it makes a difference whether one finds oneself standing on Nob Hill or Mt. Whitney. Finding a local maximum is usually easy: walk uphill until there is no place to walk. Finding the global maximum, on the other hand, is usually exceedingly complex unless the terrain has very special properties (no local maxima). The world of economic affairs is replete with local maxima. It is quite easy to devise systems in which each subsystem is optimally adapted to the other subsystems around it, but in which the equilibrium is only local, and quite inferior to distant equilibria that cannot be reached by the up-hill climb of evolution.

The Myopia of Evolution

Darwinian evolution is completely myopic. At each incremental step the evolving organism becomes fitter relative to its current environment, but there is no reason for the progress to lead to a global maximum of fitness of the individuals, separately or severally. If we are considering this kind of system, whose environment has a multitude of local maxima, we cannot understand the system unless we know something of the method and history of its evolution. Nor is there any reasonable sense in which such a system can be regarded as "fittest."

This is not just an in-principle objection to confounding hill climbing with optimization. In a myopic hill-climbing system, it may be difficult or impossible to move from a local maximum to another that is in view across a deep valley. The movement from the English system of measures to the metric system is a case in point. A society starting from scratch, and familiar with both systems, would surely prefer the metric to the English system. But if future benefits are discounted at some rate of interest, it might never be economical to switch from the one system, once adopted, to the other.

Hence, from the fact that an economic system is evolving, one cannot conclude that it has reached or is likely to reach a position that bears any resemblance to the equilibria found in the theory of perfect competition. Each species in the ecosystem is adapting to an environment of other species evolving simultaneously with it. The evolution and future of such systems can only be understood from a knowledge of their histories.

The Mechanisms of Economic Evolution

If the adaptation of both the business firm and biological species to their respective environments are instances of heuristic search, hence of local optimization or satisficing, we still have to account for the mechanisms that bring the adaptation about. In biology the mechanism is located in the genes and their success in reproducing themselves. What is the gene's counterpart in the business firm?

Nelson and Winter suggest that business firms accomplish most of their work through standard operating procedures—algorithms for making daily decisions that become routinized and are handed down from one generation of executives and employees to the next.[26] Evolution derives from all the processes that produce innovation and change in these algorithms. The fitness test is the profitability and growth rate of the firm. Profitable firms grow by the reinvestment of their profits and their attractiveness for new investment.

Nelson and Winter observe that in economic evolution, in contrast to biological evolution, successful algorithms may be borrowed by one firm from another. Thus the hypothesized system is Lamarkian, because any new idea can be incorporated in operating procedures as soon as its success is observed, and hence successful mutations can be transferred between firms. Transfer is of course not costless, but involves learning costs for the adopting firm. It may also be impeded by patent protection and commercial secrecy. Nevertheless, processes of the kinds just described play a large role in the gradual evolution of an economic system composed of business firms.

From these considerations, one sees that the evolution of firms and of economies does not lead to any easily predictable equilibrium, much less an optimum, but is a complex process, probably continuing indefinitely, that is probably best understood through an examination of its history. As in any dynamic system that has propensities for following diverging paths from almost identical starting points, equilibrium theories of an economy can tell us little about either its present state or its future.

26. R. R. Nelson and S. G. Winter, *An Evolutionary Theory of Economic Change* (Cambridge: Harvard University Press, 1982).

Human Society

Economics has been unfairly labeled the "gloomy" science, for in its Ricardian form, incorporating Malthusian views of the pressure of population on resources, it did not hold out much hope for human progress. The label is unfair, because economics in fact draws a romantic, almost heroic, picture of the human mind. Classical economics depicts humankind, individually and collectively, as solving immensely complex problems of optimizing the allocation of resources. The artfulness of the economic actors enables them to make the very best adaptations in their environments to their wants and needs. In this chapter, while keeping the adaptive capabilities of mind in the center of things, I have tried to suggest a more complex state of affairs. A veridical picture of economic actors and institutions must incorporate the information processing limits set by their inner environments. The picture must also accommodate both the conscious rationality of economic decision makers and the unplanned but adaptive evolutionary processes that have molded economic institutions.

Operations research and artificial intelligence have enhanced the procedural rationality of economic actors, helping them to make better decisions. On a larger scale, markets and organizations are social schemes that facilitate coordinated behavior, at the same time conserving the critical scarce resource of human ability to handle complexity and great masses of information. In this chapter I have not tried to evaluate these forms of individual and social organization, but simply to describe them as commonly used solutions to the central human problem of accommodating to our bounded rationality.

The analysis shows that a deeper understanding of the tools of procedural rationality requires a closer examination of how the human mind works, of the limits on human rationality. The next two chapters will describe what has been learned in the past half century about human information processing. Chapter 3 will focus on problem solving processes and general cognitive architecture, chapter 4 on memory and learning processes.

3

The Psychology of Thinking: Embedding Artifice in Nature

We watch an ant make his laborious way across a wind- and wave-molded beach. He moves ahead, angles to the right to ease his climb up a steep dunelet, detours around a pebble, stops for a moment to exchange information with a compatriot. Thus he makes his weaving, halting way back to his home. So as not to anthropomorphize about his purposes, I sketch the path on a piece of paper. It is a sequence of irregular, angular segments—not quite a random walk, for it has an underlying sense of direction, of aiming toward a goal.

I show the unlabeled sketch to a friend. Whose path is it? An expert skier, perhaps, slaloming down a steep and somewhat rocky slope. Or a sloop, beating upwind in a channel dotted with islands or shoals. Perhaps it is a path in a more abstract space: the course of search of a student seeking the proof of a theorem in geometry.

Whoever made the path, and in whatever space, why is it not straight; why does it not aim directly from its starting point to its goal? In the case of the ant (and for that matter the others) we know the answer. He has a general sense of where home lies, but he cannot foresee all the obstacles between. He must adapt his course repeatedly to the difficulties he encounters and often detour uncrossable barriers. His horizons are very close, so that he deals with each obstacle as he comes to it; he probes for ways around or over it, without much thought for future obstacles. It is easy to trap him into deep detours.

Viewed as a geometric figure, the ant's path is irregular, complex, hard to describe. But its complexity is really a complexity in the surface of the beach, not a complexity in the ant. On that same beach another small creature with a home at the same place as the ant might well follow a very similar path.

Many years ago Grey Walter built an electromechanical "turtle," having only tactile sense of its environment but capable of exploring a room, and periodically seeking its nest to recharge its batteries.[1] Today, robots with modest visual sensory capabilities roam about in a number of artificial intelligence alaboratories.[2] Suppose we undertook to design an automaton with the approximate dimensions of an ant, similar means of locomotion, and comparable sensory acuity. Suppose we provided it with a few simple adaptive capabilities: when faced with a steep slope, try climbing it obliquely; when faced with an insuperable obstacle, try detouring; and so on. (Except for problems of miniaturization of components, the present state of the art would readily support such a design.) How different would its behavior be from the behavior of the ant?

These speculations suggest a hypothesis, one that could as well have been derived as corollary from our previous discussion of artificial objects:

An ant, viewed as a behaving system, is quite simple. The apparent complexity of its behavior over time is largely a reflection of the complexity of the environment in which it finds itself.

We may find this hypothesis initially plausible or implausible. It is an empirical hypothesis, to be tested by seeing whether attributing quite simple properties to the ant's adaptive system will permit us to account for its behavior in the given or similar environments. For the reasons developed at length in the first chapter, the truth or falsity of the hypothesis should be independent of whether ants, viewed more microscopically, are simple or complex systems. At the level of cells or molecules ants are demonstrably complex, but these microscopic details of the inner environment may be largely irrelevant to the ant's behavior in relation to the outer

1. W. Grey Walter, "An Imitation of Life," *Scientific American, 185*(1950):42.

2. See, for example, R. Brooks, "A Robust-layered Control System for a Mobile Robot," *IEEE Journal of Robotics and Automation, RA-2*(1986):14–23. And a motor vehicle, NAVLAB, steered itself in the Summer of 1995 on public highways from Washington, D.C., to San Diego, California, and has also demonstrated strong capabilities for off-road navigation.

environment. That is why an automaton, though completely different at the microscopic level, might nevertheless simulate the ant's gross behavior.

In this chapter I should like to explore this hypothesis but with the word "human being" substituted for "ant."

Human beings, viewed as behaving systems, are quite simple. The apparent complexity of our behavior over time is largely a reflection of the complexity of the environment in which we find ourselves.

Now I should like to hedge my bets a little. Instead of trying to consider the "whole person," fully equipped with glands and viscera, I should like to limit the discussion to Homo sapiens, "thinking person." I myself believe that the hypothesis holds even for the whole person, but it may be more prudent to divide the difficulties at the outset, and analyze only cognition rather than behavior in general.[3]

I should also like to hedge my bets in a second way, for a human being can store away in memory a great furniture of information that can be evoked by appropriate stimuli. Hence I would like to view this information-packed memory less as part of the organism than as part of the environment to which it adapts.

The reasons for assigning some a priori probability to the hypothesis of simplicity have already been set forth in the last two chapters. A thinking human being is an adaptive system; men's goals define the interface between their inner and outer environments, including in the latter their memory stores. To the extent that they are effectively adaptive, their behavior will reflect characteristics largely of the outer environment (in the light of their goals) and will reveal only a few limiting properties of the inner environment—of the physiological machinery that enables a person to think.

3. I have sketched an extension of this hypothesis to phenomena of emotion and motivation in "Motivational and Emotional Controls of Cognition," *Psychological Review,* 74(1967):29–39, and to certain aspects of perception in "An Information-Processing Explanation of Some Perceptual Phenomena," *British Journal of Psychology,* 58(1967):1–12. Both papers are reprinted in my *Models of Thought,* vol. 1 (1979), chapters 1.3 and 6.1. The discussion of these issues is continued in "Bottleneck of Attention: Connecting Thought with Motivation," in W. D. Spaulding (ed.), *Integrative Views of Motivation, Cognition and Emotion.* Lincoln, NE: University of Nebraska Press, 1994.

I do not intend to repeat this theoretical argument at length, but rather I want to seek empirical verification for it in the realm of human thought processes. Specifically I should like to point to evidence that there are only a few "intrinsic" characteristics of the inner environment of thinking beings that limit the adaptation of thought to the shape of the problem environment. All else in thinking and problem-solving behavior is artificial—is learned and is subject to improvement through the invention of improved designs and their storage in memory.

Psychology as a Science of the Artificial

Problem solving is often described as a search through a vast maze of possibilities, a maze that describes the environment. Successful problem solving involves searching the maze selectively and reducing it to manageable proportions. Let us take, by way of specific example, a puzzle of the kind known as cryptarithmetic problems:[4]

$$\begin{array}{r} DONALD \\ + GERALD \\ \hline ROBERT \end{array} \qquad D = 5$$

The task is to replace the letters in this array by numerals, from zero through nine, so that all instances of the same letter are replaced by the same numeral, different letters are replaced by different numerals, and the resulting numerical array is a correctly worked out problem in arithmetic. As an additional hint for this particular problem, the letter D is to be replaced by the numeral 5.

One way of viewing this task is to consider all the 10!, ten factorial, ways in which ten numerals can be assigned to ten letters. The number 10! is not so large as to strike awe in the heart of a modern computer; it is only a little more than 3 million (3,628,800, to be exact). A program designed to generate all possible assignments systematically, and requiring

4. The cryptarithmetic task was first used for research on problem solving by F. Bartlett in his *Thinking* (New York: Basic Books, 1958). In the present account I have drawn on his work and on my research with Allen Newell reported in our book, *Human Problem Solving* (Englewood Cliffs, N.J.: Prentice-Hall, 1972), chapters 8–10.

a tenth of a second to generate and test each, would require at most about ten hours to do the job. (With the cue $D = 5$, only an hour would be needed.) I haven't written the program, but a tenth of a second is far longer than a computer would need to examine each possibility.

There is no evidence that a human being could do this. It might take a man as long as a minute to generate and test each assignment, and he would have great difficulty in keeping track of where he was and what assignments he had already tried. He could use paper and pencil to assist him on the latter score, but that would slow him down even more. The task, performed in this way, might call for several man-years of work—I assume a forty-hour week.

Notice that in excluding exhaustive, systematic search as a possible way for a human to solve the problem, we are making only very gross assumptions about human capabilities. We are assuming that simple arithmetic operations take times that are of the order of seconds, that the operations are essentially executed serially, rather than in parallel, and that large amounts of memory are not available in which new information can be stored at split-second speeds. These assumptions say something, but not very much, about the physiology of the human central nervous system. For example, modifying the brain by incorporating in it a new subsystem with all the properties of a desk calculator would be a quite remarkable feat of brain surgery—or evolution. But even such a radical alteration would change the relevant assumptions only slightly for purposes of explaining or predicting behavior in this problem environment.

Human beings do frequently solve the $DONALD + GERALD = ROBERT$ problem. How do they do it? What are the alternative ways of representing the environment and conducting the search?

Search Strategies

One way to cut down the search drastically is to make the assignments systematically, as before, but to assign numerals to the letters one by one so that inconsistencies can be detected before an assignment is complete, and hence whole classes of possible assignments can be ruled out at one step. Let me illustrate how this works.

Suppose we start from the right, trying assignments successively for the letters $D, T, L, R, A, E, N, B, O$, and G, and substituting numerals in the order 1, 2, 3, 4, 5, 6, 7, 8, 9, 0. We already know that $D = 5$, so we strike 5 from the list of available numerals. We now try $T = 1$. Checking in the right-hand column, we detect a contradiction, for $D + D = T + c$, where c is 10 or 0. Hence, since $(D = 5, T = 1)$ is not feasible, we can rule out all the remaining 8! assignments of the eight remaining numerals to the eight remaining letters. In the same way all possible assignments for T, except $T = 0$, can be ruled out without considering the assignments for the remaining letters.

The scheme can be improved further by the expedient of calculating directly, by addition, what assignment should be made to the sum of a column whenever the two addends are known. With this improvement we shall not need to search for the assignment for T, for $T = 0$ can be inferred directly from $D = 5$. Using this scheme, the $DONALD + GERALD = ROBERT$ problem can be solved quite readily, with paper and pencil. Ten minutes should suffice. Figure 3 shows the search tree, in slightly simplified form. Each branch is carried to the point where a contradiction is detected. For example, after the assignments $(D = 5, T = 0)$, the assignment $L = 1$ leads to the inference $R = 3$, which yields a contradiction since from the left-hand column of the problem array $R = 3$ would imply that G is negative.

Figure 3 is oversimplified in one respect. Each of the branches that terminates with a contradiction after assignment of a value to E should actually be branched one step further. For the contradiction in these cases arises from observing that no assignment for the letter O is now consistent. In each case four assignments must be examined to determine this. Thus the full search tree would have 68 branches—still a far cry from 10! or even 9!.

An enormous space has been cut down to a quite small space by some relatively small departures from systematic, exhaustive search. It must be confessed that the departures are not all as simple as I have made them appear. One step in the proposed scheme requires finding the contradictions implied by an assignment. This means of course the "relatively direct" contradictions, for if we had a rapid process capable of detecting *all* inconsistent implications, direct or indirect, it would find the problem solution almost at once. In this problem any set of assignments other than the single correct one implies a contradiction.

$D = 5$ $T = 0$ $L = 1$ $R = 3$ $G < 0$ □
$L = 2$ $R = 5$ $G = 0$ □
$L = 3$ $R = 7$ $A = 1$ $E = 2$ □
$A = 2$ $E = 4$ □
$A = 4$ $E = 8$ □
$A = 6$ $E = 2$ □
$A = 8$ $E = 6$ □
$A = 9$ $E = 8$ □
$L = 4$ $R = 9$ $A = 1$ $E = 2$ □
$A = 2$ $E = 4$ □
$A = 3$ $E = 6$ □
$A = 6$ $E = 2$ □
$A = 7$ $E = 4$ □
$A = 8$ $E = 6$ □
$L = 6$ $R = 3$ $G < 0$ □
$L = 7$ $R = 5$ □
$L = 8$ $R = 7$ $A = 1$ $E = 3$ □
$A = 2$ $E = 5$ □
$A = 3$ $E = 7$ □
$A = 4$ $E = 9$ $N = 1$ $B = 8$ □
$N = 2$ $B = 9$ □
$N = 3$ $G = 0$ □
$N = 6$ $O = 2$ $G = 1$

Figure 3
Possible search tree for $DONALD + GERALD = ROBERT$

What is meant by searching for direct contradictions is something like this: after a new assignment has been made, those columns are examined where the newly substituted letter occurs. Each such column is solved, if possible, for a still-unassigned letter, and the solution checked to see whether this numeral remains unassigned. If not, there is a contradiction.

In place of brute-force search we have now substituted a combined system of search and "reason." Can we carry this process further; can we eliminate substantially all trial-and-error search from the solution method? It turns out that we can for this problem, although not for all cryptarithmetic problems.[5]

The basic idea that permits us to eliminate most trial-and-error search in solving the problem before us is to depart from the systematic right-to-left assignment of numerals. Instead we search for columns of the

5. For example, the method to be described does not eliminate as much search from the cryptarithmetic problem $CROSS + ROADS = DANGER$.

problem array that are sufficiently determinate to allow us to make new assignments, or at least new inferences about the properties of assignments.

Let me go through the process briefly. From $D = 5$, we immediately infer $T = 0$, as before. We also infer that 1 is carried into the second column, hence that $R = 2L + 1$ is odd. On the extreme left, from $D = 5$, we infer that R is greater than 5 (for $R = 5 + G$). Putting together these two inferences, we have $R = 7$ or $R = 9$, but we do not try these assignments. Now we discover that the second column from the left has the peculiar structure $O + E = O$—a number plus another equals itself (apart from what is carried into or out of the column). Mathematical knowledge, or experiment, tells us that this can be true only if $E = 0$ or $E = 9$. Since we already have $T = 0$, it follows that $E = 9$. This eliminates one of the alternatives for R, so $R = 7$.

Since $E = 9$, it follows that $A = 4$, and there must be a one carried into the third column from the right; hence $2L + 1 = 17$, or $L = 8$. All that remains now is to assign 1, 2, 3, and 6 in some order to N, B, O, and G. We get $G = 1$ by observing that for any assignment of O there is a number carried into the leftmost column. We are now left with only $3! = 6$ possibilities, which we may be willing to eliminate by trial and error: $N = 6$, $B = 3$, and therefore $O = 2$.

We have traced a solution path through the problem maze on three different assumptions about the search strategy. The more sophisticated, in a certain sense, that strategy became, the less search was required. But it is important to notice that, once the strategy was selected, the course of the search depended only on the structure of the problem, not on any characteristics of the problem solver. By watching a person, or an automaton, perform in this problem environment, what could we learn about him? We might well be able to infer what strategy was followed. By the mistakes made, and the success in recovering from them, we might be able to detect certain limits of the capacity or accuracy of the individual's memory and elementary processes. We might learn something about the speed of these processes. Under favorable circumstances, we might be able to learn which among the thinkable strategies the individual was able actually to acquire and under what circumstances likely to acquire them. We should certainly be unlikely to learn anything specific about the neurological characteristics of the central nervous system, nor would the specifics of that system be relevant to his behavior, beyond placing bounds on the possible.

The Limits on Performance

Let us undertake to state in positive fashion just what we think these bounds and limits are, as revealed by behavior in problem situations like this one. In doing so, we shall draw upon both experimental evidence and evidence derived from computer simulations of human performance. The evidence refers to a variety of cognitive tasks, ranging from relatively complex ones (cryptarithmetic, chess, theorem proving), through an intermediate one (concept attainment), to simple ones that have been favorites of the psychological laboratory (rote verbal learning, short-term memory span). It is important that with this great variety of performance only a small number of limits on the adaptability of the inner system reveal themselves—and these are essentially the same limits over all the tasks. Thus the statement of what these limits are purports to provide a single, consistent explanation of human performance over this whole range of heterogeneous task environments.

Limits on Speed of Concept Attainment

Extensive psychological research has been carried out on concept attainment within the following general paradigm.[6] The stimuli are a set of cards bearing simple geometric designs that vary, from card to card, along a number of dimensions: shape (square, triangle, circle), color, size, position of figure on card, and so on. A "concept" is defined extensionally by some set of cards—the cards that are instances of that concept. The concept is defined intensionally by a property that all the instances have in common but that is not possessed by any of the remaining cards.

6. This account of concept attainment is based on the paper with my late colleague Lee Gregg, "Process Models and Stochastic Theories of Simple Concept Formation," *Journal of Mathematical Psychology,* 4(June 1967):246–276. See also A. Newell and H. A. Simon, "Overview: Memory and Process in Concept Formation," chapter 11 in B. Kleinmuntz (ed.), *Concepts and the Structure of Memory* (New York: Wiley, 1967), pp. 241–262. The former paper is reprinted in *Models of Thought,* vol. 1, chapter 5.4.

Examples of concepts are "yellow" or "square" (simple concepts), "green triangle" or "large, red" (conjunctive concepts), "small or yellow" (disjunctive concept), and so on.

In our discussion here I shall refer to experiments using an N-dimensional stimulus, with two possible values on each dimension, and with a single relevant dimension (simple concepts). On each trial an instance (positive or negative) is presented to the subject, who responds "Positive" or "Negative" and is reinforced by "Right" or "Wrong," as the case may be. In typical experiments of this kind, the subject's behavior is reported in terms of number of trials or number of erroneous responses before an error-free performance is attained. Some, but not all, experiments ask the subject also to report periodically the intensional concept (if any) being used as a basis for the responses.

The situation is so simple that, as in the cryptarithmetic problem, we can estimate a priori how many trials, on the average, a subject should need to discover the intended concept provided that the subject used the most efficient discovery strategy. On each trial, regardless of response, the subject can determine from the experimenter's reinforcement whether the stimulus was actually an instance of the concept or not. If it was an instance, the subject knows that one of the attribute values of the stimulus—its color, size, shape, for example—defines the concept. If it was not an instance, the subject knows that the *complement* of one of its attribute values defines the concept. In either case each trial rules out half of the possible simple concepts; and in a random sequence of stimuli each new stimulus rules out, on the average, approximately half of the concepts not previously eliminated. Hence the average number of trials required to find the right concept will vary with the logarithm of the number of dimensions in the stimulus.

If sufficient time were allowed for each trial (a minute, say, to be generous), and if the subject were provided with paper and pencil, any subject of normal intelligence could be taught to follow this most efficient strategy and would do so without much difficulty. As these experiments are actually run, subjects are not instructed in an efficient strategy, are not provided with paper and pencil, and take only a short time—typically four seconds, say—to respond to each successive stimulus. They also use many more trials to discover the correct concept than the number calculated from the efficient strategy. Although the experiment has not, to my knowledge, been run, it is fairly certain that, even with training, a subject who was required to respond in four seconds and not allowed paper and pencil would be unable to apply the efficient strategy.

What do these experiments tell us about human thinking? First, they tell us that human beings do not always discover for themselves clever strategies that they could readily be taught (watching a chess master play a duffer should also convince us of that). This is hardly a very startling conclusion, although it may be an instructive one. I shall return to it in a moment.

Second, the experiments tell us that human beings do not have sufficient means for storing information in memory to enable them to apply the efficient strategy unless the presentation of stimuli is greatly slowed down or the subjects are permitted external memory aids, or both. Since we know from other evidence that human beings have virtually unlimited semipermanent storage (as indicated by their ability to continue to store odd facts in memory over most of a lifetime), the bottleneck in the experiment must lie in the small amount of rapid-access storage (so-called short-term memory) available and the time required to move items from the limited short-term store to the large-scale long-term store.[7]

From evidence obtained in other experiments, it has been estimated that only some seven items can be held in the fast, short-term memory and that perhaps as many as five to ten seconds are required to transfer an item from the short-term to the long-term store. To make these statements operational, we shall have to be more precise, presently, about the meaning of "item." For the moment let us assume that a simple concept is an item.

Even without paper and pencil a subject might be expected to apply the efficient strategy if (1) he was instructed in the efficient strategy and

7. The monograph by J. S. Bruner, J. J. Goodnow, and G. A. Austin, *A Study of Thinking* (New York: Wiley, 1956) was perhaps the first work to emphasize the role of short-term memory limits (their term was "cognitive strain") in performance on concept-attainment tasks. That work also provided rather definite descriptions of some of the subjects' strategies.

(2) he was allowed twenty or thirty seconds to respond to and process the stimulus on each trial. Since I have not run the experiment, this assertion stands as a prediction by which the theory may be tested.

Again the outcome may appear obvious to you, if not trivial. If so, I remind you that it is obvious only if you accept my general hypothesis: that in large part human goal-directed behavior simply reflects the shape of the environment in which it takes place; only a gross knowledge of the characteristics of the human information-processing system is needed to predict it. In this experiment the relevant characteristics appear to be (1) the capacity of short-term memory, measured in terms of number of items (or "chunks," as I shall call them); (2) the time required to fixate an item, or chunk, in long-term memory. In the next section I shall inquire as to how consistent these characteristics appear to be over a range of task environments. Before I do so, I want to make a concluding comment about subjects' knowledge of strategies and the effects of training subjects.

That strategies can be learned is hardly a surprising fact, nor that learned strategies can vastly alter performance and enhance its effectiveness. All educational institutions are erected on these premises. Their full implication has not always been drawn by psychologists who conduct experiments in cognition. Insofar as behavior is a function of learned technique rather than "innate" characteristics of the human information-processing system, our knowledge of behavior must be regarded as sociological in nature rather than psychological—that is, as revealing what human beings in fact learn when they grow up in a particular social environment. When and how they learn particular things may be a difficult question, but we must not confuse learned strategies with built-in properties of the underlying biological system.

The data that have been gathered, by Bartlett and in our own laboratory, on the cryptarithmetic task illustrate the same point. Different subjects do indeed apply different strategies in that task—both the whole range of strategies I sketched in the previous section and others as well. How they learned these, or how they discover them while performing the task, we do not fully know (see chapter 4), although we know that the sophistication of the strategy varies directly with a subject's previous exposure to and comfort with mathematics. But apart from the strategies the only human characteristic that exhibits itself strongly in the crypt-arithmetic task is the limited size of short-term memory. Most of the difficulties the subjects have in executing the more combinatorial strategies (and perhaps their general aversion to these strategies also) stem from the stress that such strategies place on short-term memory. Subjects get into trouble simply because they forget where they are, what assignments they have made previously, and what assumptions are implicit in assignments they have made conditionally. All of these difficulties would necessarily arise in a processor that could hold only a few chunks in short-term memory and that required more time than was available to transfer them to long-term memory.

The Parameters of Memory—Eight Seconds per Chunk

If a few parameters of the sort we have been discussing are the main limits of the inner system that reveal themselves in human cognitive behavior, then it becomes an important task for experimental psychology to estimate the values of these parameters and to determine how variable or constant they are among different subjects and over different tasks.

Apart from some areas of sensory psychology, the typical experimental paradigms in psychology are concerned with hypothesis testing rather than parameter estimating. In the reports of experiments one can find many assertions that a particular parameter value is—or is not—"significantly different" from another but very little comment on the values themselves. As a matter of fact the pernicious practice is sometimes followed of reporting significance levels, or results of the analysis of variance, without reporting at all the numerical values of the parameters that underlie these inferences.

While I am objecting to publication practices in experimental psychology, I shall add another complaint. Typically little care is taken in choosing measures of behavior that are the most relevant to theory. Thus in learning experiments "rate of learning" is reported, almost indifferently, in terms of "number of *trials* to criterion," "total number of *errors*," "total *time* to criterion," and perhaps other measures as well. Specifically the practice of reporting learning rates in terms of trials rather than time, prevalent through the first half of this century, and almost up to the present time, not only hid from view the remarkable constancy of the parameter I am about to discuss but also led to much meaningless dispute over "one-trial" versus "incremental" learning.[8]

Ebbinghaus knew better. In his classic experiments on learning nonsense syllables, with himself as subject, he recorded both the number of repetitions and the amount of time required to learn sequences of syllables of different length. If you take the trouble to calculate it, you find that the *time per syllable* in his experiments works out to about ten to twelve seconds.[9]

I see no point in computing the figure to two decimal places—or even to one. The constancy here is a constancy to an order of magnitude, or perhaps to a factor of two—more nearly comparable to the constancy of the daily temperature, which in most places stays between 263° and 333° Kelvin, than to the constancy of the speed of light. There is no reason to be disdainful of a constancy to a factor of two. Newton's original estimates of the speed of sound contained a fudge factor of 30 per cent (eliminated only a hundred years later), and today some of the newer physical "constants" for elementary particles are even more vague. Beneath any approximate, even very rough, constancy, we can usually expect to find a genuine parameter whose value can be defined accurately once we know what conditions we must control during measurement.

If the constancy simply reflected a parameter of Ebbinghaus—albeit one that held steady over several years—it would be more interesting to biography than psychology. But that is not the case. When we examine some of the Hull-Hovland experiments of the 1930s, as reported, for ex-

8. The evidence of the constancy of the fixation parameter is reviewed in L. W. Gregg and H. A. Simon, "An Information-Processing Explanation of One-Trial and Incremental Learning," *Journal of Verbal Learning and Verbal Behavior,* 6(1967):780–787; H. A. Simon and E. A. Feigenbaum, "An Information-Processing Theory of Verbal Learning," *ibid.,* 3(1964):385–396; Feigenbaum and Simon, "A Theory of the Serial Position Effect," *British Journal of Psychology,* 53(1962):307–320; E. A. Feigenbaum, "An Information-Processing Theory of Verbal Learning," unpublished doctoral dissertation, Pittsburgh: Carnegie Institute of Technology, 1959; and references cited therein. All these papers save the last are reprinted in *Models of Thought,* vol. 1

9. Herman Ebbinghaus, *Memory* (New York: Dover Publications, 1964), translated from the German edition of 1885, especially pp. 35–36, 40, 51.

ample, in Carl Hovland's chapter in S. S. Stevens's *Handbook,* we find again (after we calculate them, for trials are reported instead of times) times in the neighborhood of ten or fifteen seconds for college sophomores to fixate nonsense syllables of low meaningfulness by the serial anticipation method. When the drum speed increases (say from four seconds per syllable to two seconds per syllable), the number of trials to criterion increase proportionately, but the total learning time remains essentially constant.

There is a great deal of gold in these hills. If past nonsense-syllable experiments are re-examined from this point of view, many are revealed where the basic learning parameter is in the neighborhood of fifteen seconds per *syllable.* You can make the calculation yourself from the experiments reported, for example in J. A. McGeoch's *Psychology of Human Learning.* B. R. Bugelski, however, seems to have been the first to make this parameter constancy a matter of public record and to have run experiments with the direct aim of establishing it.[10]

I have tried not to exaggerate how constant is "constant." On the other hand, efforts to purify the parameter measurement have hardly begun. We do know about several variables that have a major effect on the value, and we have a theoretical explanation of these effects that thus far has held up well.

We know that meaningfulness is a variable of great importance. Nonsense syllables of high association value and unrelated one-syllable words are learned in about one-third the time required for nonsense syllables of low association value. Continuous prose is learned in about one-third the time per word required for sequences of unrelated words. (We can get the latter figure also from Ebbinghaus' experiments in memorizing *Don Juan.* The times *per symbol* are roughly 10 percent of the corresponding times for nonsense syllables.)

We know that similarity—particularly similarity among stimuli—has an effect on the fixation parameter somewhat less than the effect of meaningfulness, and we can also estimate its magnitude on theoretical grounds.

10. B. R. Bugelski, "Presentation Time, Total Time, and Mediation in Paired-Associate Learning," *Journal of Experimental Psychology,* 63(1962):409–412.

The theory that has been most successful in explaining these and other phenomena reported in the literature on rote verbal learning is an information-processing theory, programmed as a computer simulation of human behavior, dubbed EPAM.[11] Since EPAM has been reported at length in the literature, I shall not discuss it here, except for one point that is relevant to our analysis. The EPAM theory gives us a basis for understanding what a "chunk" is. A chunk is a maximal familiar substructure of the stimulus. Thus a nonsense syllable like "QUV" consists of the chunks "Q," "U," "V"; but the word "CAT" consists of a single chunk, since it is a highly familiar unit. EPAM postulates constancy in the time required to fixate a chunk. Empirically the constant appears to be about eight seconds per chunk, or perhaps a little more. Virtually all the quantitative predictions that EPAM makes about the effects of meaningfulness, familiarity, and similarity upon learning speed follow from this conception of the chunk and of the constancy of the time required to fixate a single chunk.

In fixation of new information, EPAM first adds new branches to its discrimination net then adds information to images at terminal nodes of the branches. There is growing evidence that the eight seconds for fixation in long-term memory is required only for expanding the net, and that information can be added in a second or two to locations (variable-places) in images that are already present in an expert's long-term memory. Such images are called retrieval structures or templates. We will return to this point in discussing expert memory. EPAM's architecture and memory processes are described in H. B. Richman, J. J. Staszewski and H. A. Simon, "Simulation of Expert Memory Using EPAM IV," *Psychological Review, 102*(1995):305–330.

The Parameters of Memory—Seven Chunks, or Is It Two?

The second limiting property of the inner system that shows up again and again in learning and problem-solving experiments is the amount of

11. For a survey of the range of phenomena for which EPAM has been tested, see E. A. Feigenbaum and H. A. Simon, "EPAM-like Models of Recognition and Learning," *Cognitive Science, 8*(1984): 305–336, reprinted in *Models of Thought,* vol. 2 (1989), chapter 3.4.

information that can be held in short-term memory. Here again the relevant unit appears to be the chunk, where this term has exactly the same meaning as in the definition of the fixation constant.

Attention was attracted to this parameter, known previously from digit-span, numerosity-judging, and discrimination tasks, by George Miller's justly celebrated paper on "The Magical Number Seven, Plus or Minus Two."[12] It is no longer as plausible as it was when he wrote his paper that a single parameter is involved in the three kinds of task, rather than three different parameters: we shall consider here only tasks of the digit-span variety. Today we would express the parameter as the amount of information that can be rehearsed in about two seconds, which is, in fact, about seven syllables or short words.

The facts that appear to emerge from recent experiments on short-term memory are these. If asked to read a string of digits or letters and simply to repeat them back, a subject can generally perform correctly on strings up to seven or even ten items in length. If almost any other task, however simple, is interposed between the subject's hearing the items and repeating them, the number retained drops to two. From their familiarity in daily life we could dub these numbers the "telephone directory constants." We can generally retain seven numbers from directory to phone if we are not interrupted in any way—not even by our own thoughts.

Where experiments appear to show that more than two chunks are retained across an interruption, the phenomena can almost always be explained parsimoniously by mechanisms we have already discussed in the previous section. In some of these experiments the explanation—as already pointed out by Miller—is that the subject recodes the stimulus into a smaller number of chunks before storing it in short-term memory. If ten items can be recoded as two chunks, then ten items can be retained. In the other experiments where "too much" appears to be retained in short-term memory, the times allowed the subjects permit them in fact to fixate the excess of items in long-term memory. For experts who have acquired retrieval structures or templates in their domain of expertise into which the new information can be inserted, these times can be quite short—a second or two per item.

12. *Psychological Review,* 63(1956):81–97.

Putting aside expert performance for the moment, I shall cite just two examples from the literature. N. C. Waugh and D. A. Norman report experiments, their own and others', that show that only the first two of a sequence of items is retained reliably across interruption, but with some residual retention of the remaining items.[13] Computation of the fixation times available to the subjects in these experiments shows that a transfer rate to long-term memory of one chunk per five seconds would explain most of the residuals. (This explanation is entirely consistent with the theoretical model that Waugh and Norman themselves propose.)

Roger Shepard has reported that subjects shown a very long sequence of photographs—mostly landscapes—can remember which of these they have seen (when asked to choose from a large set) with high reliability.[14] When we note that the task is a recognition task, requiring storage only of differentiating cues, and that the average time per item was about six seconds, the phenomenon becomes entirely understandable—indeed predictable—within the framework of the theory that we are proposing.

The Organization of Memory

I have by no means exhausted the list of experiments I could cite in support of the fixation parameter and the short-term capacity parameter and in support of the hypothesis that these parameters are the principal, and almost only, characteristics of the information-processing system that are revealed, or could be revealed, by these standard psychological experiments.

This does not imply that there are not other parameters, and that we cannot find experiments in which they are revealed and from which they can be estimated. What it does imply is that we should not look for great complexity in the laws governing human behavior, in situations where the behavior is truly simple and only its environment is complex.

In our laboratory we have found that mental arithmetic tasks, for instance, provide a useful environment for teasing out other possible pa-

13. N. C. Waugh and D. A. Norman, "Primary Memory," *Psychological Review,* 72(1965):89–104.

14. Roger N. Shepard, "Recognition Memory for Words, Sentences, and Pictures," *Journal of Verbal Learning and Verbal Behavior,* 6(1957):156–163.

rameters. Work that Dansereau has carried forward shows that the times required for elementary arithmetic operations and for fixation of intermediate results account for only part—perhaps one-half—of the total time for performing mental multiplications of four digits by two. Much of the remaining time appears to be devoted to retrieving numbers from the memory where they have been temporarily fixated, and "placing" them in position in short-term memory where they can be operated upon.[15]

Stimulus Chunking

I should like now to point to another kind of characteristic of the inner system—more "structural" and also less quantitative—that is revealed in certain experiments. Memory is generally conceived to be organized in an "associative" fashion, but it is less clear just what that term is supposed to mean. One thing it means is revealed by McLean and Gregg. They gave subjects lists to learn—specifically 24 letters of the alphabet in scrambled order. They encouraged, or induced, chunking of the lists by presenting the letters either one at a time, or three, four, six, or eight on a single card. In all of the grouped conditions, subjects learned in about half the time required in the one-at-a-time condition.[16]

McLean and Gregg also sought to ascertain whether the learned sequence was stored in memory as a single long list or as a hierarchized list of chunks, each of which was a shorter list. They determined this by measuring how subjects grouped items temporally when they recited the list, and especially when they recited it backwards. The results were clear: the alphabets were stored as sequences of short subsequences; the subsequences tended to correspond to chunks presented by the experimenter, or sublengths of those chunks; left to his own devices, the subject tended to prefer chunks of three or four letters. (Recall the role of chunks of this length in the experiments on effects of meaningfulness in rote learning.)

15. See Donald F. Dansereau and Lee W. Gregg, "An Information Processing Analysis of Mental Multiplication," *Psychonomic Science*, 6(1966):71–72. The parameters of memory are discussed in more detail in *Models of Thought*, vol. 1, chapters 2.2, and 2.3; and vol. 2, chapter 2.4; and in Richman, Staszewski and Simon, *op. cit.*

16. R. S. McLean and L. W. Gregg, "Effects of Induced Chunking on Temporal Aspects of Serial Recitation," *Journal of Experimental Psychology*, 74(1967): 455–459.

Visual Memory

The materials in the McLean-Gregg experiments were strings of symbols. We might raise similar questions regarding the form of storage of information about two-dimensional visual stimuli.[17] In what sense do memory and thinking represent the visual characteristics of stimuli? I do not wish to revive the debate on "imageless thought"—certainly not in the original form that debate took. But perhaps the issue can now be made more operational than it was at the turn of the century.

As I enter into this dangerous ground, I am comforted by the thought that even the most fervent opponents of mentalism have preceded me. I quote, for example, from B. F. Skinner's *Science and Human Behavior* (1952, p. 266):

> A man may see or hear "stimuli which are not present" on the pattern of the conditioned reflexes: he may see X, not only when X is present, but when any stimulus which has frequently accompanied X is present. The dinner bell not only makes our mouth water, it makes us see food.

I do not know exactly what Professor Skinner meant by "seeing food," but his statement gives me courage to say what an information-processing theory might mean by it. I shall describe in a simplified form one kind of experiment that has been used to throw light on the question. Suppose we allow a subject to memorize the following visual stimulus—a magic square:

4   9   2

3   5   7

8   1   6

Now we remove the stimulus and ask the subject a series of questions about it, timing his or her answers. What numeral lies to the right of 3, to the right of 1? What numeral lies just below 5? What numeral is diagonally above and to the right of 3? The questions are not all of the same difficulty—in fact I have arranged them in order of increasing difficulty

17. The letters in the stimuli of the McLean-Gregg experiment are, of course, also two-dimensional visual stimuli. Since they are familiar chunks, however, and can be immediately recognized and recoded, there is no reason to suppose that their two-dimensional character plays any role in the subject's behavior in the experiment. Again this is "obvious," but only if we already have a general theory of how stimuli are processed "inside."

and would expect a subject to take substantially longer to answer the last question than the first.

Why should this be? If the image stored in memory were isomorphic to a photograph of the stimulus, we should expect no large differences in the times required to answer the different questions. We must conclude that the stored image is organized quite differently from a photograph. An alternative hypothesis is that it is a list structure—a hypothesis that is consistent, for example, with the data from the McLean-Gregg experiment and that is much in the spirit of information-processing models of cognition.

For example, if what was stored were a list of lists: "TOP," "MIDDLE," "BOTTOM," where "TOP" is 4–9–2, "MIDDLE" is 3–5–7, and "BOTTOM" is 8–1–6; the empirical results would be easy to understand. The question "What numeral lies to the right of 3?" is answered by searching down lists. The question "What numeral lies just below 5?" is answered, on the other hand, by matching two lists, item by item—a far more complex process than the previous one.

There is no doubt, of course, that a subject could *learn* the up–down relations or the diagonal relations as well as the left–right relations. An EPAM-like theory would predict that it would take the subject about twice as long to learn both left-right and up-down relations as the former alone. This hypothesis can be easily tested, but, to the best of my knowledge, it has not been.

Evidence about the nature of the storage of "visual" images, pointing in the same direction as the example I have just given, is provided by the well-known experiments of A. de Groot and others on chess perception.[18] De Groot put chess positions—taken from actual games—before subjects for, say, five seconds; then he removed the positions and asked the subjects to reconstruct them. Chess grandmasters and masters could reconstruct the positions (with perhaps 20 to 24 pieces on the board) almost without error, while duffers were able to locate hardly any of the pieces correctly, and the performance of players of intermediate skill fell somewhere

18. Adriaan D. de Groot, "Perception and Memory versus Thought: Some Old Ideas and Recent Findings," in B. Kleinmuntz (ed.), *Problem Solving* (New York: Wiley, 1966), pp. 19–50. See also the work by Chase and Simon reported in chapters 6.4 and 6.5 of *Models of Thought* vol. 1.

between masters and duffers. But the remarkable fact was that, when masters and grandmasters were shown other chessboards with the same numbers of pieces *arranged at random,* their abilities to reconstruct the boards were only marginally better than the duffers' with the boards from actual games, while the duffers performed as well or poorly as they had before.

What conclusion shall we draw from the experiment? The data are inconsistent with the hypothesis that the chess masters have some special gift of visual imagery—or else why the deterioration of their performance? What the data suggest strongly is that the information about the board is stored in the form of *relations* among the pieces, rather than a "television scan" of the 64 squares. It is inconsistent with the parameters proposed earlier—seven chunks in short-term memory and five seconds to fixate a chunk—to suppose that anyone, even a grandmaster, can store 64 pieces of information (or 24) in ten seconds. It is quite plausible that he can store (in short-term and long-term memory) information about enough relations (supposing each one to be a familiar chunk) to permit him to reproduce the board of figure 4:

1. Black has castled on the K's side, with a fianchettoed K's bishop defending the K's Knight.
2. White has castled on the Q's side, with his Queen standing just before his King.
3. A Black pawn on his K5 and a White pawn on his Q5 are attacked and defended by their respective K's and Q's Knights, the White Queen also attacking the Black pawn on the diagonal.
4. White's Q-Bishop attacks the Knight from KN5.
5. The Black Queen attacks the White K's position from her QN3.
6. A Black pawn stands on its QB4.
7. A White pawn on K3 blocks that advance of the opposing Black pawn.
8. Each side has lost a pawn and a Knight.
9. White's K-Bishop stands on K2.

Pieces not mentioned are assumed to be in their starting positions. Since some of the relations as listed are complex, I shall have to provide reasons for considering them unitary "chunks." I think most strong chess players would regard them as such. Incidentally I wrote down these relations from my own memory of the position, in the order in which they

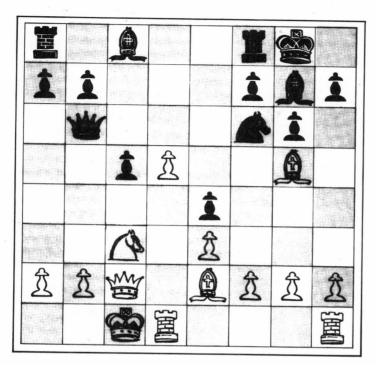

Figure 4
Chess position used in memory experiment

occurred to me. Eye-movement data for an expert chess player looking at this position tend to support this analysis of how the relations are analyzed and stored.[19] The eye-movement data exhibit with especial clarity the relations 3 and 5.

The expert can store the information about the position even more rapidly if he or she recognizes the standard opening to which it belongs—in this case, the Gruenfeld Defense—thereby accessing a familiar template that gives the positions of about a dozen pieces.

The implication of this discussion of visual memory for my main theme is that many of the phenomena of visualization do not depend in any

19. O. K. Tikhomirov and E. D. Poznyanskaya, "An Investigation of Visual Search as a Means of Analysing Heuristics," English translation from *Voprosy psikhologii*, 1966, vol. 12, in *Soviet Psychology*, 2(Winter 1966–1967):3–15. See also *Models of Thought*, vol. 1, chapters 6.2 and 6.3.

detailed way upon underlying neurology but can be explained and predicted on the basis of quite general and abstract features of the organization of memory—features which are essentially the same ones that were postulated in order to build information-processing theories of rote learning and of concept attainment phenomena.

Specifically, we are led to the hypothesis that memory is an organization of list structures (lists whose components can also be lists), which include descriptive components (two-termed relations) and short (three-element or four-element) component lists. A memory with this form of organization appears to have the right properties to explain storage phenomena in both visual and auditory modalities, and of pictorial and diagrammatic as well as propositional (verbal and mathematical) information.

The Mind's Eye

The experiments we have been discussing relate not only to visual long-term memory, but also to the Mind's Eye, the short-term memory where we hold and process mental images. In the mind's eye we can often substitute "seeing" for reasoning. Consider the economist's common supply-and-demand diagram, which shows, by one curve, the quantity of a commodity that will be supplied to the market at each price, and by another curve, the quantity that will be demanded at each price. If we notice that the two curves intersect, we can interpret the intersection as the point at which the supply and demand quantities are equal, a point of market equilibrium; and we can read off directly from the x-axis and y-axis of the diagram the equilibrium quantity and price (the x and y coordinates of the intersection). All this processing goes on in the mind's eye, using the information read from the diagram.

Alternatively, we could write down the equations for the two lines and solve them simultaneously to find the same equilibrium quantity and price. Using visual processes and algebraic ones we attain the same knowledge, but by completely different computational paths (and perhaps with vastly different amounts of labor and insight). In many scientific fields, inferences are made with a combination of verbal, mathematical and diagrammatic reasoning—certain inferences being reached more easily in one form, others in another. In Alfred Marshall's famous textbook, *Principles of Economics*, the text is wholly verbal, the diagrams are provided in footnotes, and the corresponding algebra is given in a mathematical appendix, thus allowing readers full freedom to adopt their preferred representation in each instance.

To understand the interplay of these and other modes of human inference, we need to study the computational processes required to reach conclusions in each representation. Currently, this is a very active area of cognitive research.[20]

Processing Natural Language

A theory of human thinking cannot and should not avoid reference to that most characteristic cognitive skill of human beings—the use of language. How does language fit into the general picture of cognitive processes that I have been sketching and into my general thesis that psychology is a science of the artificial?

Historically the modern theory of transformational linguistics and the information-processing theory of cognition were born in the same matrix—the matrix of ideas produced by the development of the modern digital computer, and in the realization that, though the computer was embodied in hardware, its soul was a program. One of the initial professional papers on transformational linguistics and one of the initial professional papers on information-processing psychology were presented, the one after the other, at a meeting at MIT in September 1956.[21] Thus the two bodies of theory have had cordial relations from an early date, and quite rightly, for they rest conceptually on the same view of the human mind.

20. J. Larkin and H. A. Simon, "Why a Diagram is (Sometimes) Worth 10,000 Words," *Cognitive Science*, 11(1987):65–100; A. M. Leonardo, H. J. M. Tabachneck and H. A. Simon, "A Computational Model of Diagram Reading and Reasoning," *Proceedings of the 17th Annual Conference of the Cognitive Science Society* (1995); Y. Qin and H. A. Simon, "Imagery and Mental Models of Problem Solving," in J. Glasgow, N. H. Narayanan and B. Chandrasekaran (eds.), *Diagrammatic Reasoning: Computational and Cognitive Perspectives* (Menlo Park, CA: AAAI/The MIT Press, 1995).

21. N. Chomsky, "Three Models for the Description of Language," and A. Newell and H. A. Simon, "The Logic Theory Machine," both in *IRE Transactions on Information Theory*, IT-2, no. 3 (September 1956).

Now some may object that this is not correct and that they rest on almost diametrically opposed views of the human mind. For I have stressed the artificial character of human thinking—how it adapts itself, through individual learning and social transmission of knowledge, to the requirements of the task environment. The leading exponents of the formal linguistic theories, on the other hand, have taken what is sometimes called a "nativist" position. They have argued that a child could never acquire any skill so complex as speaking and understanding language if he did not already have built into him at birth the basic machinery for the exercise of these skills.

The issue is reminiscent of the debate on language universals—on whether there are some common characteristics shared by all known tongues. We know that the commonalities among languages are not in any sense specific but that they relate instead to very broad structural characteristics that all languages seem to share in some manner. Something like the distinction between noun and verb—between object and action or relation—appears to be present in all human languages. All languages appear to have the boxes-within-boxes character called phrase structure. All languages appear to derive certain strings from others by transformation.[22]

Now if we accept these as typical of the universals to which the nativist argument appeals, there are still at least two different possible interpretations of that argument. The one is that the language competence is *purely* linguistic, that language is *sui generis,* and that the human faculties it calls upon are not all employed also in other performances.

An alternative interpretation of the nativist position is that producing utterances and understanding the utterances of others depend on some characteristics of the human central nervous system which are common in all languages but also essential to other aspects of human thinking besides speech and listening.

The former interpretation does not, but the latter does, provide an explanation for the remarkable parallelism holding between the underlying

22. On language universals see Joseph H. Greenberg (ed.), *Universals of Language* (Cambridge: The MIT Press, 1963), particularly Greenberg's own chapter, pp. 58–90. On the "nativist" position, see Jerrold J. Katz, *The Philosophy of Language* (New York: Harper and Row, 1966), pp. 240–282.

assumptions about human capabilities embedded in modern linguistic theory and the assumptions embedded in information-processing theories of human thinking. The kinds of assumptions that I made earlier about the structure of human memory are just the kinds of assumptions one would want to make for a processing system capable of handling language. Indeed there has been extensive borrowing back and forth between the two fields. Both postulate hierarchically organized list structures as a basic principle of memory organization. Both are concerned with how a serially operating processor can convert strings of symbols into list structures or list structures into strings. In both fields the same general classes of computer-programming languages have proved convenient for modeling and simulating the phenomena.

Semantics in Language Processing

Let me suggest one way in which the relation between linguistic theories and information-processing theories of thinking is going to be even closer in the future than it was in the past. Linguistic theory has thus far been largely a theory of syntax, of grammar. In practical application to such tasks as automatic translation, it has encountered difficulties when translation depended on more than syntactic cues—when it depended on context and meaning. It seems pretty clear that one of the major directions that progress in linguistics will have to take is toward development of an adequate semantics to complement syntax.

The theory of thinking I have been outlining can already provide an important part of such a semantic component. The principles of memory organization I have described can be used as a basis for discussing the internal representation of both linguistic strings and two-dimensional visual stimuli, or other nonlinguistic stimuli. Given these comparable bases for the organization of the several kinds of stimuli, it becomes easier to conceptualize the cooperation of syntactic and semantic cues in the interpretation of language.

Several research projects have been carried out at Carnegie Mellon University that bear on this point. I should like to mention just two of these, which illustrate how this approach might be used to explain the resolution of syntactic ambiguities by use of semantic cues.

L. Stephen Coles, in a dissertation completed in 1967, described a computer program that uses pictures on a cathode ray tube to resolve syntactic ambiguities.[23] I shall paraphrase his procedure with an example that is easier to visualize than any he actually used. Consider the sentence:

I saw the man on the hill with the telescope.

This sentence has at least three acceptable interpretations; a linguist could, no doubt, discover others. Which of the three obvious ones we pick depends on where we think the telescope is: Do I have it? Does the man on the hill have it? Or is it simply on the hill, not in his hands?

Now suppose that the sentence is accompanied by figure 5. The issue is no longer in doubt. Clearly it is I who have the telescope.

Coles's program is capable of recognizing objects in a picture and relations among objects; and it is capable of representing the picture as a list structure, which, in the example before us, we might describe thus:

SAW ((I, WITH (telescope)), (man, ON (hill))).

I have not tried to reproduce the actual details of the scheme he used, but I have simply shown that a picture, so represented, could readily be matched against alternate parsings of a verbal string and thus used to resolve the ambiguity of the latter.

Another program, completed by Laurent Siklóssy, illustrates how semantic information can aid in the *acquisition* of a language.[24] The reader may be familiar with the "Language through Pictures" books developed by Professor I. A. Richards and his associates. These books have been prepared for a large number of languages. On each page is a picture and beneath it one or more sentences that say something about the picture in the language to be learned. The sequence of pictures and accompanying sentences is arranged to proceed from very simple situations ("I am here," "That is a man") to more complex ones ("The book is on the shelf").

Siklóssy's program takes as its input an analogue to one of the "Language through Pictures" books. The picture is assumed to have already

23. L. Stephen Coles, *Syntax Directed Interpretation of Natural Language,* doctoral dissertation, Carnegie Institute of Technology, 1967. A slightly abridged version is reprinted in H. A. Simon and L. Siklóssy (eds.), *Representation and Meaning* (Englewood Cliffs, N.J.: Prentice-Hall, 1972).

24. Also reprinted in *Representation and Meaning.*

Figure 5
A syntactically ambiguous sentence; "I saw the man on the hill with the telescope"

been transformed into a list structure (not unlike the one illustrated earlier for Coles's system) as its internal representation. The program's task is to learn, when confronted with such a picture, to utter the appropriate sentence in the natural language it is learning—a sentence that says what the picture shows. In the case of the sentence about the telescope (somewhat more complicated than any on which the scheme has actually been tested), one would hope that the program would respond to the picture with "I saw the man on the hill with the telescope," if it were learning English, or *Ich habe den Mann auf dem Berg mit dem Fernglas gesehen,* if it were learning German.

Of course the program could respond correctly only if it had learned earlier, in the context of other sentences, the lexical and syntactical components required for the translation. A child trying to understand the sentence must meet the same requirement. In other cases the program would use the sentence associated with the picture to add to its vocabulary and syntax.[25]

25. I may mention in passing that Siklóssy's system refutes John Searle's notorious "Chinese Room Paradox," which purports to prove that a computer cannot understand language. As Siklóssy's program shows, if the room has windows on the world (which Searle's room doesn't) the system matches words, phrases and sentences to their meanings by comparing sentences with the scenes they denote.

I do not wish to expand some pioneering experiments into a comprehensive theory of semantics. The point of these examples is that they show that the kind of memory structure that has been postulated, for other reasons, to explain human behavior in simpler cognitive tasks is suitable for explaining how linguistic strings might be represented internally, how other kinds of stimuli might be similarly represented, and how the communalities in representation—the use of hierarchically organized list structures for both—may explain how language and "meanings" come together in the human head.

There is no contradiction, then, between the thesis that a human being possesses, at birth, a competence for acquiring and using language and the thesis that language is the most artificial, hence also the most human of all human constructions. The former thesis is an assertion that there *is* an inner environment and that it does place limits on the kinds of information processing of which the organism is capable. The structure of language reveals these limits; and these limits in turn account for such commonality as exists among the Babel of human tongues.

The latter thesis, of the artificiality of language, is an assertion that the limits on adaptation, on possible languages, imposed by the inner environment are very broad limits on organization, not very specific limits on syntax. Moreover, according to the thesis, they are limits imposed not only on language but also on every other mode of representing internally experience received through stimuli from outside.

Such a view of the relation of language and thinking puts a new cast on the "Whorfian" hypothesis that—stating it in overstrong form—only the expressible is thinkable. If the view is valid, it would be as correct to say. "Only the thinkable is expressible"—a view that, I suppose, Kant would have found quite congenial.

Conclusion

The thesis with which I began this chapter was the following:

Human beings viewed as behaving systems, are quite simple. The apparent complexity of our behavior over time is largely a reflection of the complexity of the environment in which we find ourselves.

That hypothesis was based in turn on the thesis of the first chapter: that behavior is adapted to goals, hence is artificial, hence reveals only those characteristics of the behaving system that limit the adaptation.

To illustrate how we have begun to test these theses and at the same time to build up a theory of the simple principles that underlie human behavior, I have surveyed some of the evidence from a range of human performances, particularly those that have been studied in the psychological laboratory.

The behavior of human subjects in solving cryptarithmetic problems, in attaining concepts, in memorizing, in holding information in short-term memory, in processing visual stimuli, and in performing tasks that use natural languages provides strong support for these theses. The artificiality—hence variability—of human behavior hardly calls for evidence beyond our observation of everyday life. The experiments are therefore mostly significant in what they show about the broad commonalities in organizations of the human information-processing system as it engages in different tasks.

The evidence is overwhelming that the system is basically serial in its operation: that it can process only a few symbols at a time and that the symbols being processed must be held in special, limited memory structures whose content can be changed rapidly. The most striking limits on subjects' capacities to employ efficient strategies arise from the very small capacity of the short-term memory structure (seven chunks) and from the relatively long time (eight seconds) required to transfer a chunk of information from short-term to long-term memory.

The claim that the human cognitive system is basically serial has been challenged in recent years by advocates of neural nets and parallel connectionist models of the nervous system. I would make the following cautionary observations. Although there is clearly a lot of parallelism in the sensory organs (especially eyes and ears), after stimuli have been recognized seriality is enforced by the small capacity of the short-term memory that is employed in the subsequent stages of processing. There is also a moderate degree of parallelism in the processing of motor signals, but again, only after the initial signals have passed through the STM bottleneck. Third, seriality of processing at the symbolic level, the level with which we are concerned here, says nothing, one way or the other, about the extent of seriality or parallelism in the neural implementation of the symbolic processing at the next level below. (By an ironic reverse twist, parallel connectionist networks are routinely simulated by programs run on serial computers of standard von Neumann architecture.)

Finally, a large part of the discernible parallel neural activity in the brain may well consist only in passive maintenance of memory, the active processes being largely localized and serial. (Evidence now coming from magnetic resonance imaging [MRI] of the brain is consistent with this view.) The speeds at which people can perform cognitive tasks and the usual limits on the numbers of tasks they can perform concurrently do not provide much evidence for (or need for) parallel processing capacity. Until connectionism has demonstrated, which it has not yet done, that complex thinking and problem-solving processes can be modeled as well with parallel connectionist architectures as they have been with serial architectures, and that the experimentally observed limits on concurrent cognitive activity can be represented in the connectionist models, the case for massive parallelism outside the sensory functions remains dubious.

When we turn from tasks that exercise mainly the short-term memory and serial-processing capabilities of the central nervous system to tasks that involve retrieval of stored information, we encounter new limits of adaptation, and through these limits we acquire new information about the organization of mind and brain. Studies of visual perception and of tasks requiring use of natural language show with growing clarity that memory is indeed organized in associative fashion, but that the "associations" have the properties of what, in the computer trade, are usually called "list structures." I have indicated briefly what those properties are, and more will be said about them in the next chapter.

These are the sorts of generalizations about human thinking that are emerging from the experimental evidence. They are simple things, just as our hypothesis led us to expect. Moreover, though the picture will continue to be enlarged and clarified, we should not expect it to become essentially more complex. Only human pride argues that the apparent intricacies of our path stem from a quite different source than the intricacy of the ant's path.

One of the curious consequences of my approach—of my thesis—is that I have said almost nothing about physiology. But the mind is usually thought to be located in the brain. I have discussed the organization of the mind without saying anything about the structure of the brain.

The main reason for this disembodiment of mind is of course the thesis that I have just been discussing. The difference between the hardware of a computer and the "hardware" of the brain has not prevented computers from simulating a wide spectrum of kinds of human thinking—just because both computer and brain, when engaged in thought, are adaptive systems, seeking to mold themselves to the shape of the task environment.

It would be unfortunate if this conclusion were altered to read that neurophysiology has nothing to contribute to the explanation of human behavior. That would be of course a ridiculous doctrine. But our analysis of the artificial leads us to a particular view of the form that the physiological explanation of behavior must take. Neurophysiology is the study of the inner environment of the adaptive system called Homo sapiens. It is to physiology that we must turn for an explanation of the limits of adaptation: Why is short-term memory limited to seven chunks; what is the physiological structure that corresponds to a "chunk"; what goes on during the eight seconds that a chunk is being fixated; how are associational structures realized in the brain?

As our knowledge increases, the relation between physiological and information-processing explanations will become just like the relation between quantum-mechanical and physiological explanations in biology (or the relation between solid-state physics and programming explanations in computer science). They constitute two linked levels of explanation with (in the case before us) the limiting properties of the inner system showing up at the interface between them.

Finally, we may expect also that, as we link information-processing psychology to physiology on the inner side, we shall also be linking psychology to the general theory of search through large combinatorial spaces on the outer side—the side of the task environment. But that is the topic of my fifth chapter, for the theory of design *is* that general theory of search. Before we take up that topic we must say more about how the large bodies of information used by designers are stored in the human mind and accessed.

4

Remembering and Learning: Memory as Environment for Thought

In developing the proposition in chapter 3 that human thought processes are simple, the cards were perhaps stacked by the examples selected to illustrate the thesis. A task like $DONALD + GERALD = ROBERT$ is difficult enough for an intelligent adult, but it does not call on much information stored in one's memory. The solver must know the numbers, how to add and subtract them, and perhaps a few facts about parity, but that is about all. Contrast this with the task of driving a taxi in Pittsburgh or in the East Bay. No amount of intelligence will take the cab driver from here to there unless he has stored in memory an enormous amount of information about the names of streets, their locations, and mutual intersections. (The street index of my Pittsburgh atlas contains about 8,500 entries.) If this information is available in memory, however, choosing a route probably does not call for a very complex strategy.[1]

The hypothesis that human thought processes are simple emerged from the information-processing research of the 1950s and 1960s. Most of that research employed puzzlelike tasks, similar to the cryptarithmetic problems and concept attainment tasks discussed in the last chapter, which could be performed without great dependence on memory or skills previously learned. Additional examples are the Missionaries and Cannibals puzzle, the Tower of Hanoi puzzle, and problems of logical inference, all of which have been studied extensively in the psychological laboratory

---

1. I believe this is true, but it is not obvious. Exercise for the reader: write a computer program that, given the street map of an area and some knowledge of which streets are trunk routes, will choose a reasonable path to deliver a passenger from one point to another.

and support the picture of human thinking that was drawn in the last chapter.

It is reasonable that research on human thinking should begin with relatively contentless tasks of these kinds but not that it should end there. And so in recent decades research in both cognitive psychology and artificial intelligence has been turning more and more to semantically rich domains—domains that have substantial, meaningful content, where skillful performance calls upon large amounts of specialized knowledge retrieved from memory. Does human thinking still look simple in such domains?

In pursuing this question, we shall be interested especially in high-level performance of the kinds of tasks that confront professionals in their everyday work or college students who are preparing for professional practice. Among the professional-level domains that have been studied fairly extensively in the laboratory, and hence some of whose parameters are known, are chess playing, making medical diagnoses, solving college physics problems, and discovering regularities in empirical data. We will use these and others as examples.

Except for chessplaying, long-term memory played only a modest role in performance of the tasks examined in chapter 3. The simplicity we discovered there was largely a simplicity of process (only a few basic symbol manipulating processes had to be postulated to account for the behavior) and a simplicity of the architecture of the mind (its seriality and its limited short-term memory). A few parameters, especially the chunk capacity of STM and the storage time for new chunks in LTM, played a dominant role in fixing the limits of the system's performance.

As we move to semantically rich domains, new questions of simplicity and complexity arise. Does the richness of the contents of long-term memory imply complexity of structure, or can that richness be accommodated by the simple organizations of list structures that were described briefly in chapter 3? Is a higher level of complexity required for programs that exploit these large stores of memory, or are the same processes in evidence as those that account for problem solving in the puzzlelike tasks of chapter 3? Do the learning programs required to store new data and processes in long-term memory introduce new levels of complexity? We will see that the evidence from studies of human performance and from its simulation by computer generally supports the hypothesis of simplicity. More memory does not necessarily mean more complexity.

Semantically Rich Domains

There is a certain arbitrariness in drawing the boundary between inner and outer environments of artificial systems. In our discussion of economic behavior in chapter 2, we might well have considered the business firm's cost function to be part of the inner environment. Instead we abstracted the decision-making process from the production technology and regarded only the limits on rational calculation as inner constraints on adaptivity. The cost function was treated, along with the demand function, as part of the outer environment to which the firm was seeking to adapt.

We can adopt a similar viewpoint toward the human problem solver, whose basic tool for solving problems is a small repertory of information processes of the sorts described in the last chapter. This processor operates on an outer environment that has two major components: the "real world," sensed through eye, ear, and touch, and acted upon by leg, hand, and tongue, and a large store of (correct and incorrect) information about that world, held in long-term memory and retrievable by recognition or by association. When the processor is solving puzzlelike problems, the memory plays a limited role. The structure of the problem rather than the organization of memory steers the problem-solving search. When it is solving problems in semantically rich domains, a large part of the problem-solving search takes place in long-term memory and is guided by information discovered in that memory. Hence an explanation of problem solving in such domains must rest on an adequate theory of memory.

Long-Term Memory

Certain facts about human long-term memory (LTM) were set forth in the last chapter. It is of essentially unlimited size—no one appears ever to have been able to fill his memory to overflowing, although in senility new items cannot be stored. About eight seconds are required to store a new chunk in LTM, except when an expert has an already stored template with which the chunk can be associated, in which case only a second or two is required."[2] A rather shorter time (a few hundred milliseconds to a couple of seconds) is needed to retrieve information previously stored. The memory is usually described as "associative" because of the way in which one thought retrieved from it leads to another. Information is stored in linked list structures.

In terms of our present-day knowledge of LTM, we can extend this description a bit. We can think of the memory as a large encyclopedia or library, the information stored by topics (nodes), liberally cross-referenced (associational links), and with an elaborate index (recognition capability) that gives direct access through multiple entries to the topics. Long-term memory operates like a second environment, parallel to the environment sensed through eyes and ears, through which the problem solver can search and to whose contents he can respond.

Medical diagnosis is a semantically rich domain that has now been investigated extensively, with the aims both of understanding the diagnostic processes used by physicians and of building systems for diagnosis by computer. The thickness of medical textbooks and reference books attests to the large volume of information that is required for accurate diagnosis. When the diagnostic strategies of physicians are studied, two kinds of processes are prominent in their thinking-aloud protocols: processes of direct *recognition*, where presence of a symptom leads almost immediately to hypothesizing a disease that might be its cause, and processes of *search* quite like those identified in the simpler problem-solving tasks that were described in chapter 3.[3] The diagnosis generally proceeds from symptoms to hypothesized disease entities, to tests for resolving doubts and weeding out alternatives, to new symptoms, and so on. Thus the search is conducted alternately in each of two environments: the physician's mental library of medical knowledge and the patient's body. Information gleaned from one environment is used to guide the next step of search in the other.

2. But see the discussion of retrieval structures in chapter 3.

3. Arthur Elstein et al., *Medical Problem Solving* (Cambridge, Mass.: Harvard University Press, 1978). There is now on the market a fully automated diagnostic system for internal medicine, *Doctor's Assistant,* that is based largely on this model of the diagnostic process and that has performed well in clinical trials.

Intuition

What about the sudden flashes of "intuition" that sometimes allow the expert to arrive immediately at the answer that the novice can find (if at all) only after protracted search? (I put "intuition" in quotes to emphasize that it is a label for a process, not an explanation of it.) Intuition is a genuine enough phenomenon which can be explained rather simply: most intuitive leaps are acts of recognition. Let me illustrate this with the game of chess.

In chapter 3 I described the remarkable ability of chess masters and grandmasters to reproduce chess positions almost faultlessly after seeing them for five or ten seconds. This performance was reconciled with the known limits of short-term memory by observing that for the chess master a position from a game does not consist of 25 isolated pieces but of five or six chunks, each one a familiar configuration that may be a template of a dozen pieces or a smaller chunk consisting of two to five or more related pieces. Since we can estimate, at least roughly, the amount of variety in chess positions from well-played games, we can also estimate the number of familiar chunks that must be stored in the master's long-term memory to make his performance possible. Several different methods of estimation all lead to numbers of the general magnitude of 50,000. We need not take the exact number seriously, but it is interesting that it is of the same order of magnitude as the natural language recognition vocabulary of a college-educated reader.[4]

Hence we can say that one part of the grandmaster's chess skill resides in the 50,000 chunks stored in memory, and in the index (in the form of a structure of feature tests) that allows him to recognize any one of these chunks on the chess board and to access the information in long-term memory that is associated with it. The information associated with familiar patterns may include knowledge about what to do when the pattern is encountered. Thus the experienced chess player who recognizes the feature called an *open file* thinks immediately of the possibility of moving a rook to that file. The move may or may not be the best one, but it is one that should be considered whenever an open file is present. The expert

4. See Simon, *Models of Thought*, vol. 1, chapters 6.2 and 6.3.

recognizes not only the situation in which he finds himself, but also what action might be appropriate for dealing with it.

A feature-testing system capable of discriminating among 50,000 different items might make its discriminations quite rapidly. Even if each test were dichotomous, which they probably are not, only about 16 tests would have to be performed to achieve each recognition. (The game of "twenty questions" is based on the fact that 20 dichotomous tests will discriminate among a million items.) If each test required 10 milliseconds, the whole process could be performed in less than 200 milliseconds—well within the time limits of human recognition capabilities.

When playing a "rapid transit" game, at ten seconds a move, or fifty opponents simultaneously, going rapidly from one board to the next, a chess master is operating mostly "intuitively," that is, by recognizing board features and the moves that they suggest. The master will not play as well as in a tournament, where about three minutes, on the average, can be devoted to each move, but nonetheless will play relatively strong chess. A person's skill may decline from grandmaster level to the level of a master, or from master to expert, but it will by no means vanish. Hence recognition capabilities, and the information associated with the patterns that can be recognized, constitute a very large component of chess skill.[5]

How Much Information?

The amount of information stored by chess masters appears to be roughly consonant with the amount of information that professionals in other

---

5. Strong empirical evidence for this claim will be found in F. Gobet and H. A. Simon, "The Roles of Recognition Processes and Look-Ahead Search in Time-Constrained Expert Problem Solving: Evidence from Grandmaster Level Chess," *Psychological Science*, 7(1)(January 1996): 52–55. My colleague Hans Berliner has built a powerful backgammon program, which has beaten the world's champion human player in a match, using pattern recognition capabilities rather than search processes as the basis for its skill. (See "Computer Backgammon," *Scientific American*, 242(6)(June 1980):64–85.) By contrast, most extant computer chess programs carry out enormous searches for lack of strong recognition capabilities. For a survey of the current status of computer chess programs see H. A. Simon and J. Schaeffer, "The Game of Chess," in R. J. Aumann and S. Hart (eds.), *Handbook of Game Theory*, vol. 1, (Netherlands: Elsevier, 1992).

domains possess—although only the roughest measures of these quantities are available. At first it might appear unlikely that disciplines as disparate as chess, medicine, mathematics, and chemistry should call on memory stores of comparable size. But their comparability says little about the nature of the domains. No one knows everything there is to know about chess, medicine, chemistry, or any other serious domain. Here, as elsewhere, man must be the measure of skill. A professional's knowledge is adequate when he or she knows about as much as other professionals in the same discipline. What places an upper limit on professional knowledge is the amount of time that can be devoted to acquiring and maintaining it—some fraction of a human waking lifetime.

From what is known about the rates at which people can store new information in long-term memory, 50,000 chunks is not an unconscionable amount of knowledge to acquire in a decade, say, of professional training. Of course 50,000 chunks is an underestimate of what the chess master (or other professional) knows, but even if we raise the estimate by one or two orders of magnitude, that much information could probably be acquired in a decade. If it takes thirty seconds of attention to store a new chunk in long-term memory (8 seconds for initial acquisition, say, plus 22 seconds to overlearn for permanent retention), then ten years of intensive study at 1,500 hours per year (about four hours per day) could produce a memory store of 1.8 million chunks. Even a dedicated professional who worked around the clock with a minimum of daydreaming would be unlikely to learn more than that, for a substantial part of the time would probably be spent not in learning but in practicing what had already been learned.

In a couple of domains where the matter has been studied, we do know that even the most talented people require approximately a decade to reach top professional proficiency. Except for Bobby Fisher and Judit Polgar, who reached grandmaster status in nine years and some months from the time they first began to play chess, there is no record of anyone achieving that level in less than a decade. Unless we except Mozart, there is no record of a composer producing first-rate music before he had completed a decade of serious study and practice; and even in the case of Mozart the music that he composed between the seventh and tenth years after he began writing is notable as Mozart juvenalia rather than the music of a "grandmaster."[6]

When a domain reaches a point where the knowledge for skillful professional practice cannot be acquired in a decade, more or less, then several adaptive developments are likely to occur. Specialization will usually increase (as it has, for example, in medicine), and practitioners will make increasing use of books and other external reference aids in their work.

Architecture is a good example of a domain where much of the information a professional requires is stored in reference works, such as catalogues of available building materials, equipment, and components, and official building codes. No architect expects to keep all of this in his head or to design without frequent resort to these information sources. In fact architecture can almost be taken as a prototype for the process of design in a semantically rich task domain. The emerging design is itself incorporated in a set of external memory structures: sketches, floorplans, drawings of utility systems, and so on. At each stage in the design process, the partial design reflected in these documents serves as a major stimulus for suggesting to the designer what he should attend to next. This direction to new subgoals permits in turn new information to be extracted from memory and reference sources and another step to be taken toward the development of the design.[7] I will have a little more to say about this cycle of design activities, and its implications for style, in the next chapter.

It should not be supposed that every advance in human knowledge increases the amount of information that has to be mastered by professionals. On the contrary, some of the most important progress in science is the discovery and testing of powerful new theories that allow large numbers of facts to be subsumed under a few general principles. There is a constant competition between the elaboration of knowledge and its compression into more parsimonious form by theories. Hence it is not safe to say that the professional chemist must learn more today than a half century ago, before the general laws of quantum mechanics were announced.

6. The information about composing was compiled by my colleague, John R. Hayes (personal communication). From preliminary data he has also gathered, it appears that similar statements can be made about painting.

7. Ömer Akin, *Psychology of Architectural Design* (London: Pion Limited, 1986).

It is probably safe to say that the chemist must know as much as a diligent person can learn in about a decade of study.

Memory for Processes
Memory has been discussed here as though it consisted mainly of a body of *data*. But experts possess skills as well as knowledge. They acquire not only the ability to recognize situations or to provide information about them; they also acquire powerful special skills for dealing with situations as they encounter them. Physicians prescribe and operate as well as diagnose.

The boundary between knowledge and skill is subtle. For example, when we write a computer program in any language except machine language, we are really not writing down processes but data structures. These data structures are then *interpreted* or *compiled* into processes— that is, into machine-language instructions that the computer can understand and execute. Nevertheless for most purposes it is convenient for us simply to ignore the translation step and to treat the computer programs in higher-level languages as representing processes.

We can think of a medical diagnostic system (human or computer) as having a large body of medical knowledge, together with a few general processes for drawing inferences from it. Or we can think of the knowledge as organized in processes, instructing the expert how to proceed with the diagnosis, for example:

If you find that the patient has a high fever, then test for the following additional symptoms.

Similarly, a student's knowledge of geometry could be stored as theorems:

If two triangles have the three pairs of corresponding sides equal, then they are congruent.

or, alternatively, as condition-action pairs (called *productions*):

Test the corresponding sides of two triangles for pairwise equality; if all are equal, store the assertion that the triangles are congruent.

Whether expertness is stored as data or process, or some combination of both, does not alter what we have said about complexity. The specialized knowledge and skill can still be regarded as residing in the external environment of long-term memory, to be drawn upon by general processes that control and steer problem-solving search—processes like means-ends analysis and recognition that we have already identified in the simpler task environments discussed in chapter 3.

Understanding and Representation

Efforts to solve a problem must be preceded by efforts to understand it. Here is an example of a puzzle-like task that most people find reasonably difficult:

A Tea Ceremony

In the inns of certain Himalayan villages is practiced a most civilized and refined tea ceremony. The ceremony involves a host and exactly two guests, neither more nor less. When his guests have arrived and have seated themselves at his table, the host performs five services for them. These services are listed below in the (increasing) order of the nobility which the Himalayans attribute to them.

Stoking the Fire
Fanning the Flames
Passing the Rice Cakes
Pouring the Tea
Reciting Poetry

During the ceremony, any of those present may ask another, "Honored Sir, may I perform this onerous task for you?" However, a person may request of another only the least noble of the tasks the other is performing. Further, if a person is performing any tasks, then he may not request a task which is nobler than the least noble task he is already performing. Custom requires that by the time the tea ceremony is over, all the tasks will have been transmitted from the host to the most senior of the guests. How may this be accomplished?

Before a General Problem Solver (see chapter 5) can go to work on the Tea Ceremony problem, it has to extract from the written statement a description of the problem in terms of constructs that a GPS can deal with: symbol structures, tests for differences between structures, operators that alter structures, and symbolized goals and tests for their achievement. A GPS understands a problem when the problem has been presented to it in terms of such entities, so that its processes for detecting differences, finding relevant operators, applying operators, and evaluating progress toward the solution can go into action.

Now the Tea Ceremony problem really has nothing to do with inns in Himalayan villages. Underneath it is an abstract problem about two classes of objects (*participants* and *tasks*), relations between objects (each task is *assigned* to a participant), an ordering of the tasks (by *nobility*), and operators (*transferring* a task from one participant to another). Understanding the problem requires extracting these entities from the natural language text.

A Program that Understands

A computer program, UNDERSTAND, simulates the processes that people use to generate an internal representation of (to understand) a problem like A Tea Ceremony.[8] UNDERSTAND proceeds in two phases: it parses the sentences of the problem instructions, and then constructs the representation from the information it has extracted from the parsed sentences.

The task of analysing natural-language sentences has already been discussed in the last chapter: it involves inferring from the linear string of words the implied hierarchic structure of phrases and clauses. The UNDERSTAND program accomplishes this in a quite orthodox way, similar to that of other extant parsing programs. The second phase (construction) is more interesting. Here, the parsed sentences are examined to discover what objects and sets of objects are being referred to, what properties of objects are mentioned and what are the relations among them, which of the predicates and relations describe *states* and which describe *moves*, and what the goal state is. UNDERSTAND then proceeds to construct a format for representing states and to generate programs for making legal moves by changing one state into another.

For example, in A Tea Ceremony a state could be represented by a list of the three participants, each described by a list of the tasks he is performing. Another list could indicate the ordering of the five tasks by nobility. The legal move program would delete a task from the list of a particular participant (the donor) and add it to the list of another (the

8. UNDERSTAND is described and its behavior discussed in chapters 7.1 to 7.3 of my *Models of Thought*, vol. 1. The program and chapters were produced jointly by John R. Hayes and myself.

donee), after checking to see that the task was not more noble than others on the donor's or donee's lists.

Since (as was argued in the last chapter) list structures have a quite general capacity for representing symbolic information of all kinds, a program like UNDERSTAND is capable, in principle, of constructing a representation for virtually any kind of puzzlelike problem that does not require real-world knowledge for its understanding; for any such problem can be described in terms of objects, their relations, and changes in their relations.[9]

Understanding Physics

In contrast with understanding a problem like A Tea Ceremony, understanding problems in domains that have rich semantics requires prior knowledge of the domain. Consider the simple statics problem:

The foot of a ladder rests against a vertical wall and on a horizontal floor. The top of the ladder is supported from the wall by a horizontal rope 30 ft long. The ladder is 50 ft long, weighs 100 lb with its center of gravity 20 ft from the foot, and a 150-lb man is 10 ft from the top.

Determine the tension in the rope.

In order to go to work on this problem, a person must know what a coefficient of friction is, that a ladder may be regarded as a lever with a fulcrum and with forces applied to it, that a man may be abstracted to a mass or a fulcrum, and many facts of a similar sort. What distinguishes this kind of problem from A Tea Ceremony is not that it has real-world reference, but that it makes reference to matters that are supposed already to be known.

Gordon Novak has written a very interesting program, ISAAC, that can understand physics (statics) problems like the one described above.[10] ISAAC is able to do this because it has stored in memory information about levers, masses, inclined planes, and the like in the form of simple schemas that describe objects of these kinds and that indicate the kinds of

9. Of course the UNDERSTAND program actually implemented is only a prototype for the engine that would be needed to accomplish this in all generality.

10. G. S. Novak, "Representation of Knowledge in a Program for Solving Physics Problems," Proceedings of the Fifth International Joint Conference on Artificial Intelligence, 1977, pp. 286–291.

information associated with them. A ladder schema, for example, looks something like this:

Ladder
Type: ladder
Locations: (of foot, top, other points mentioned)
Supports:
Length:
Weight:
Attachments: (to other objects)

When a problem is presented to ISAAC, it begins, like UNDERSTAND, to parse the sentences of the problem statement. In ISAAC's case, however, more is involved than identifying objects and relations and representing them appropriately. Particular kinds of objects whose meanings are already known (i.e., provided with schemas in ISAAC's memory) must be recognized and identified with their schemas, and the "slots" in the schemas must be filled in with the requisite information. A ladder must be recognized as a lever, and a copy must be constructed of the lever schema specifying the length of the ladder, its weight, its center of gravity, the location of its fulcrum and of the forces impinging on it, and so on.

Having identified the appropriate object schemas and accumulated the appropriate information about them, ISAAC is then able to assemble the individual schemas (describing the ladder, the man, the surfaces on which the ladder rests) into a composite problem schema. Using the latter schema as guide, the program then constructs and solves the equations that are appropriate for describing the equilibria of forces.

ISAAC is prototypic of systems for understanding problems in semantically rich domains. Knowledge of physics is stored in the program in two ways: in the component schemas, which guide the process of generating a representation of the problem state (the problem schema), and in the procedures for generating the equations of equilibrium (the laws of statics which correspond to the processes for creating the operators in a program like UNDERSTAND).

When we compare the two understanding programs, we see that UNDERSTAND has to create its problem representation and operators out of whole cloth, guided only by the information in the problem instructions, while ISAAC has to discover a match between the things mentioned in the problem statement and the schemas and physical laws it has stored in memory. A more sophisticated understanding system would combine these two capabilities. One component of the system (corresponding to UNDERSTAND) would generate state representations when confronted with new problem domains and would store these as sets of schemas. The other component (corresponding to ISAAC) would endeavor to use representations already stored to interpret new problems presented to it.

However primitive the existing understanding programs may be, they do provide a set of basic mechanisms, a theory, to explain how human beings are able to grasp problems, both in new domains about which they have no knowledge and in domains about which they have a greater or lesser amount of previous semantic knowledge. The two particular systems I have described are members of a growing family of such programs that explicate the processes of understanding in an ever-widening collection of tasks. Some of these systems are concerned with relatively ill-defined tasks. For example, there has been some research on the processes that may be used for understanding children's tales or newspaper stories. Unlike the problem-solving tasks discussed here, where it is easy to test whether the system has understood the problem or not (i.e., whether it has constructed a representation that the problem solver can use to find an answer), the tests of whether a story has been "understood" are ambiguous. In the latter tasks understanding can be achieved to various degrees and various depths.

The understanding programs provide us with additional insights about visual imagery, a topic discussed in chapter 3. The state descriptions produced by UNDERSTAND and the element schemas and problem schemas of ISAAC are excellent examples of the kinds of symbolic structures proposed in that previous discussion as mental images. As a matter of fact Novak has written a subsidiary program, as part of ISAAC, that produces from the problem schema it has constructed an actual (if simple) drawing of the problem situation that can be displayed on a cathode-ray tube.

Size and Simplicity

The problem-understanding programs take information from the world outside (in these cases in the form of natural-language text) and transform it into knowledge that is stored in long-term memory as list structures or procedures. When memory has acquired a photograph, highly fragmentary and often fogged, of the external world, the problem-solving processes can carry out some of their work on this internal world instead of the outside one. This is advantageous whenever it is costly to gain access to the information in the external world.

As more knowledge is acquired about more subjects, the memory store grows, essentially without limit. Regardless of the size reached by the memory, however, it continues to be constructed of the same basic components, to be organized and indexed according to the same principles, and to be operated upon by processes having the same basic form. We may say that the system becomes more complex because it grows in size, or we may say that it remains simple since its fundamental structure does not change.

We could make exactly the same comments if we were discussing the simplicity or complexity of the Library of Congress. As the number of books increases from thousands to millions to tens of millions, the number of miles of shelves required to hold them increases correspondingly. So does the number of cards in the catalogue. But in terms of the library's architecture, the growth, however impressive, can hardly be characterized as a growth in complexity. As I shall argue in chapter 8, the transition from a single-cell to a multicell organism represents a step upward in complexity; increase in the weight of a steer or in the population of a colony of algae does not.

Human beings carry around in their heads knowledge of many domains and often even expertness in several. To the extent that the domains are distinct, as they often are, their multiplicity contributes nothing to the complexity of operating in any one of them. Studying a Greek textbook in the library is made no more difficult (and no easier) by the fact that the library also contains books on Latin, Sanskrit, and classical Chinese.

Human memory is best regarded as an extension (sometimes a large extension) of the environment in which human thought processes take place and not as an increment in the complexity of these processes. What is remarkable about the whole architecture is precisely the fact that memory enables the system to operate effectively in a wide array of different task domains using the same basic equipment that it employs to understand and solve Tea Ceremony problems or simple statics problems in physics.

Learning

The external environments of thought, both the real world and long-term memory, undergo continual change. In memory the change is adaptive. It updates the knowledge about the real world and adds new knowledge. It adds new procedures that contribute to the skills in particular task domains and improves existing procedures. A scientific theory of human thinking must take account of this process of change in the contents of memory.

If the human cognitive system is truly simple, that simplicity can only be revealed by discovering the invariants that underlie change. Among these invariants are the basic parameters of memory (parameters of the inner environment) and the general search and control processes described in chapter 3. In addition to these we may look for a basic set of processes that bring about the adaptation of long-term memory that we call learning. We may hypothesize that these learning processes are the unmoved movers that can account for the change processes in a simple and invariant way.

Learning is any change in a system that produces a more or less permanent change in its capacity for adapting to its environment. Understanding systems, especially systems capable of understanding problems in new task domains, are learning systems. So is EPAM, the system described in chapter 3 that simulates human rote verbal learning. So is Siklóssy's system, also described in that chapter, for simulating first-language learning.

Any multicomponent system can be improved in a large number of ways. Nor is there any single kind of change in the human cognitive system to which the term "learning" applies exclusively. However, the multiplicity of forms of learning need not bewilder us, for they are reducible to a few fundamental species, corresponding to the main components of the cognitive system.

Along one dimension we can distinguish between acquiring information (stored data structures) and acquiring skills (stored procedures). The

UNDERSTAND program illustrates both. The state descriptions that UNDERSTAND builds constitute new knowledge, the operators, new skills. To these categories we can add the learning of new perceptual discriminations, as exemplified by EPAM. Motor skills, although based partly on the kinds of learning already enumerated, probably also have additional components.

It is too early in the development of research on learning to attempt an exhaustive taxonomy of the kinds of learning processes that will be required to account for all the sorts of learning of which the human organism is capable, but there is reason to believe that human learning of most kinds can be explained within the framework of the symbol-processing system we have been describing.

Learning with Understanding

Every teacher knows that there is a profound difference between a student learning a lesson by rote and learning it with understanding, or meaningfully. When something has been learned by rote, it can be regurgitated more or less literally, but it cannot be used as a cognitive tool. Laboratory experiments have shown that material can usually be learned more rapidly with understanding than by rote, is retained over longer periods of time, and can be transfered better to new tasks.[11]

In spite of the great pragmatic importance of the distinction between rote and meaningful learning, the difference between them is not thoroughly understood in information processing terms. Partly it is a matter of indexing: meaningful material is indexed in such a way that it can be accessed readily when it is relevant. Partly it is a matter of redundancy: meaningful material is stored redundantly, so that if any fraction of it is forgotten, it can be reconstructed from the remainder. Partly it is a matter of representation: meaningful material is stored in the form of procedures rather than "passive" data, or if stored as data, it is represented in such a way that general problem-solving processes and other procedures can readily make use of it. All of these are aspects of understanding and meaningfulness that need further exploration.

11. George Katona, *Organizing and Memorizing* (New York: Hafner Publishing Co., 1967), chapter 4.

Production Systems

In an information-processing system that consists of data structures and programs, it has usually been easier to devise methods for adding new schemas and other data structures to the existing system than to add new programs. In the early years of AI research, artificial intelligence and simulation programs were usually organized as hierarchies of routines and subroutines. Modification of a program involved modification of one or more subroutines, a task not easily accomplished.

In the past several decades a new form of program structure has become popular: the production system.[12] What commends it, especially for building systems that learn, is the simplicity and uniformity of its structure. A *production system* is a set of arbitrarily many *productions*. Each production is a process that consists of two parts—a set of tests or *conditions* and a set of *actions*. The actions contained in a production are executed whenever the conditions of that production are satisfied. In that sense, the productions operate in complete independence of each other. Productions are usually represented by the notation:

Condition → Action,

which is reminiscent of the familiar $S{\rightarrow}R$ pairs of stimulus-response psychology. Although productions are more complex objects than $S{\rightarrow}R$ pairs, it is sometimes possible to use the latter as a metaphor for the former.

A system to simulate human cognition might be constructed with productions of two kinds: those in which the conditions are tests on the contents of short-term memory and those in which the conditions are perceptual tests on the outside world. An example of a condition of the former kind might be: "If your goal is to enter the house, open the door." Here, the *goal* of entering the house would be represented by a symbol structure in STM, and STM would be tested for the presence or absence of that structure.

An example of a perceptual production might be: "If the door is locked, use your key." Here the condition is tested in the real world (by determining if the door is locked).

12. See Newell and Simon, *Human Problem Solving*.

A system whose behavior is governed by perceptual productions is sometimes called *stimulus driven* or *data driven;* one governed by goal symbols in STM, *goal driven.* A problem solver that is mainly goal driven will give the appearance of working backward from the desired goal. One that is mainly stimulus driven will give the appearance of working forward from what it knows toward the desired goal. Of course goal-directed systems will usually employ both productions with perceptual conditions and productions with goals as conditions.

Many cognitive simulations have now been modeled as production systems. But what makes production systems especially attractive for modeling is that it is relatively easy to endow them with learning capabilities—to build so-called *adaptive production systems.* Since production systems are simply sets of productions, they can be modified by deleting productions or by inserting new ones. The consequences of such changes may or may not be adaptive, but at least there is no question of how the change is to be made.

Learning from Examples

In chapters of science and mathematics textbooks that explain new procedures, one almost always finds examples that have been worked out in detail, step by step. In an elementary algebra text, for example, we might find the following:

$9X + 17 = 6X + 23,$
$3X + 17 = 23$ (subtract $6X$ from both sides),
$3X = 6$ (subtract 17 from both sides),
$X = 2$ (divide both sides by 3).

At each step the algebraic equation is modified and a "justification" given for the modification. The process terminates when an expression is found of the form:

⟨Variable⟩ = ⟨Numeral⟩

This, and similar, equations could be solved by the following production system:

If expression has the form, ⟨variable⟩ = ⟨real number⟩ → Halt.
If expression has variable term on right side → Subtract variable term from both sides, and simplify.

If expression has numerical term on left side → Subtract numerical term from both sides, and simplify.

If variable term has coefficient other than unity → Divide both sides by coefficient.

Now a clever student who encountered the worked-out example in the text, but who had not previously acquired a procedure for solving it, could learn one in the following manner. Examining the first two steps in the example, he notices what action has been performed to transform the first line into the second. He also compares the pairs of equations and notices that the term "$6X$" has disappeared from the right side and the coefficient of X has been changed on the left. By trying the action, he discovers that it produces exactly this effect. Moreover the expression from which the "$6X$" has been removed is closer in form to the final equation than is the initial expression. He now learns a new production, by taking the feature of the initial expression that is eliminated as the condition for the action. This production is the second one in our production system. By comparing the second and third equations, he similarly infers and acquires the third production, and by comparing the third and fourth equations, the fourth production. Presumably he has already acquired the first production, which represents his understanding of what the "solution" of an algebraic equation is.

I have omitted from this account some essential details, such as how the student selects the proper degree of generalization for his productions (why "variable term" instead of "$6X$" in the condition and action of the second production?). But the simplified example conveys the general idea of how an adaptive production system can acquire new skills. This particular scheme has been devised and programmed by David Neves.[13]

A set of highly effective computer tutoring systems in geometry, algebra, and programming (LISP), employing primarily a learning-from-examples paradigm, has been developed by John Anderson and his colleagues and tested successfully in high school classrooms. In the People's Republic of China, the Psychology Institute of the Chinese Acad-

13. D. M. Neves, "A Computer Program that Learns Algebraic Procedures by Examining Examples and Working Problems in a Textbook," *Proceedings of the Second National Conference of the Canadian Society for Computational Studies of Intelligence* (1978), pp. 191–195.

emy of Sciences has developed paper and pencil materials for a full three-year middle school curriculum in algebra and geometry based upon learning from examples rather than lectures or textbook exposition. These materials are now being used successfully in several hundred schools in China. In both cases, the materials were developed by analyzing the tasks to determine what productions the students would need to acquire for effective performance and then what sequence of examples would induce the learning of these productions. These two extensive projects, in the U.S. and China, provide strong evidence for the relevance to human learning of the processes that Neves and others postulated on the basis of computer simulation.[14]

The idea of learning from examples can be extended to a method of learning "by doing." Suppose a problem-solving system is able to solve a particular problem but does it inefficiently after a great deal of search. The path to a solution finally discovered, stripped of all the extraneous branchings in the search, could serve as a worked-out example to which the procedures of the previous paragraphs could be applied. Anzai and Simon have constructed a "learning by doing" scheme of this kind for the Tower of Hanoi puzzle which, by solving the problem several times in succession, gradually acquires an efficient and general strategy.[15]

Discovery Processes

No sharp line divides learning things that are already known to others from learning things that are new to the world. What constitutes novelty depends on what knowledge is already in the mind of the problem solver and what help is received from the environment in adding to this knowledge. We should expect, therefore, that processes very similar to those employed in learning systems can be used to construct systems that discover new knowledge.

14. J. R. Anderson, A. T. Corbett, K. R. Koedinger and R. Pelletier, "Cognitive Tutors: Lessons Learned," *Journal of the Learning Sciences,* 4(1995):167–207; X. Zhu and H. A. Simon, "Learning Mathematics from Examples and by Doing," *Cognition and Instruction,* 4(1987):137–166.

15. Y. Anzai and H. A. Simon, "The Theory of Learning by Doing," *Psychological Review,* 86(1979):124–140.

Problem Solving without a Goal

Making discoveries belongs to the class of ill-structured problem-solving tasks that have relatively ill-defined goals. To discover gold, one does not even have to be looking for it (although frequently one is), and if silver or copper shows up instead of gold, that outcome will usually be welcome too. The test that something has been discovered is that something new has emerged that could not have been predicted with certainty and that the new thing has value or interest of some kind.

An early discovery program, developed in 1963, performed the letter series completion task. Given a sequence like "A B M C D M E F M," what letters come next? The answer that would be scored as correct is "G H M I J M," etc. To find the pattern, one looks for pattern in the sequence. Every third letter is M. The first letter in every triad is next in the alphabet to the second letter in the preceding triad, and the second letter in the triad is next in the alphabet to the first. The answer is simply an extrapolation of this pattern.[16] The sequence extrapolator was a harbinger of scientific law discovery programs of the 1980s and 1990s. The central idea was to use a hypothesis generator to search for pattern in data and to use indications of pattern, as detected, to guide the continuation of the search—"nothing but" our old friend, heuristic search.

Another early discovery program was AM.[17] Its task was to discover interesting new concepts and interesting conjectures about them. AM had criteria for judging what was interesting, a set of search heuristics (based on best-first search), and elementary knowledge of some task domain (e.g., elementary set theory). The program showed a considerable ability to discovery interesting new (to it) concepts, like the concept of prime number; but there was some controversy about the bases for its discoveries, some critics claiming that the new concepts were implicit in the LISP language in which AM was programmed. I do not share this dismissive view, but it would divert us from our path to discuss the issues here.

16. *Models of Thought*, vol. 1, chapters 5.1 and 5.2 (with Kenneth Kotovsky).

17. D. B. Lenat, "Automated Theory Formation in Mathematics," *Proceedings of the Fifth International Joint Conference on Artificial Intelligence* (1977), pp. 833–842. See also W. Shen, "Functional Transformation in AI Discovery Systems," *Artificial Intelligence*, 41(1989):257–272.

Rediscovering Classic Physics

Another discovery system of considerable interest is the BACON program, in its several forms,[18] which is capable of discovering invariants in bodies of numerical data. Given data on the distances of the planets from the Sun and on the periods of their orbits, it discovers that the ratio of the cubes of the periods to the squares of the distances is the same for all the planets (Kepler's third law). From data on the variation of electric current with the length of the resistance wire in a circuit, it infers Ohm's law. In a similar manner it finds the gas laws, Galileo's law of falling bodies, and many others.

BACON will introduce new concepts in order to explain the invariants it has found. Given data showing that, when two bodies accelerate each other, the ratios of the accelerations are always the same, it invents the concept of mass and associates a mass with each of the bodies. In a similar manner it invents the concept of refractive index (in Snell's law), of specific heat, and of chemical valence.

As with AM there is little that is novel in the basic construction of BACON. Given two data sets, if it finds that one datum varies monotonically (directly or inversely) with the other, it tests whether their ratio (or product) is invariant. If it succeeds, it has discovered a lawful relation in the data; if it fails, it has defined a new variable, which can then be added to the others and the process repeated. What is remarkable about the system's behavior is that finding laws of the sorts mentioned above by this procedure does not require extensive search. Seldom is it necessary to examine more than a dozen functions of the original variables in order to find an invariant.

AM and BACON have been followed by a whole host of other discovery programs that are illuminating many different aspects of scientific discovery; a number of these are discussed in *Scientific Discovery*. Not only are such systems capable of discovering new concepts but also they can plan sequences of experiments, postulate reaction paths for complex chemical reactions, induce rules for interpreting data from mass

18. P. Langley, H. A. Simon, G. L. Bradshaw and J. M. Zytkow, *Scientific Discovery: Computational Explorations of the Creative Process* (Cambridge, MA: The MIT Press, 1987).

spectrogram analysis, and enlarge the state space of a system to accommodate variables that are not directly observable.

Part of this research has been conceived as adding artificial intelligence to the tools of science. For example, the DENDRAL and MECHEM programs have generated discoveries that have been published in chemical journals as contributions to chemistry. Much of the research, however, has aimed at deepening our understanding of human discovery processes. The BACON simulations, for example, have been compared with historical cases of discovery in physics and chemistry, and some parallel experiments have been run with human subjects in the laboratory to compare their attempts at discovery with those of BACON and with those recorded in the histories of science.[19]

The AM and BACON programs and their successors give us some reason to believe that discovery processes do not introduce new kinds of complexity into human cognition. The demonstration will become more convincing when one of these systems discovers something of interest that is novel not only to it but to the world. That test has not yet been passed.

Finding New Problem Representations

Every problem-solving effort must begin with creating a representation for the problem—a problem space in which the search for the solution can take place. Of course, for most of the problems we encounter in our daily personal or professional lives, we simply retrieve from memory a representation that we have already stored and used on previous occasions. Sometimes, we have to adapt the representation a bit to the new situation, but that is usually a rather simple matter.

Occasionally, however, we encounter a situation that doesn't seem to fit any of the problem spaces we have encountered before, even with some stretching and shaping. Then we are faced with a task of discovery that may be as formidable as finding a new natural law. Newton was able to discover the law of gravitation because he had previously found a new representation, the differential calculus, that Wren and Hooke, among others who were also searching for the law, didn't have. More often, prob-

19. Y. Qin and H. A. Simon, "Laboratory Replication of Scientific Discovery Processes," *Cognitive Science, 14*(1990):281–312.

lems of representation arise that are midway in difficulty between simply adapting a known representation and inventing the calculus.

In the difficult "insight" problem known as the "Mutilated Checkerboard," a checkerboard is covered by 32 dominoes, each of which covers exactly two squares. Someone mutilates the board by snipping off the upper-left and lower-right corner squares. Can the mutilated board be covered by (31) dominoes? If so, how; if not, why not?

Typically, subjects work for hours in the given problem space of the board and dominoes to find a covering. After frustration sets in, they begin to consider changing their representation. But how? One doesn't come equipped with a "space" of representations, or even a generator of possible representations. The few successful subjects notice, at some point, that in their unsuccessful attempts at covering the board the uncovered squares are always of one color. Once they notice this, they frequently notice also that each domino covers exactly one square of each color, and the idea follows quickly that dominoes can only cover a board that has equal numbers of squares of each color, which the mutilated board does not.

Observation of subjects solving (or not solving) the mutilated checkerboard give us some important insights into the problem of discovering new representations. Focus of attention is the key to success—focusing on the particular features of the situation that are relevant to the problem, then building a problem space containing these features but omitting the irrelevant ones. This single idea falls far short of a theory of representation change but takes a first step toward building such a theory. The process of discovering new representations is a major missing link in our theories of thinking and is currently a major area of research in cognitive psychology and artificial intelligence.[20]

Conclusion

Nothing that we have discovered about memory requires us to revise our basic verdict about the complexity or simplicity of human cognition. We can still maintain that,

20. C. A. Kaplan and H. A. Simon, "In Search of Insight," *Cognitive Psychology*, 22(1990):374–419.

Human beings, viewed as behaving systems, are quite simple. The apparent complexity of our behavior over time is largely a reflection of the complexity of the environment in which we find ourselves . . .

provided that we include in what we call the human environment the cocoon of information, stored in books and in long-term memory, that we spin about ourselves.

That information, stored both as data and as procedures and richly indexed for access in the presence of appropriate stimuli, enables the simple basic information processes to draw upon a very large repertory of information and strategies, and accounts for the appearance of complexity in their behavior. The inner environment, the hardware, is simple. Complexity emerges from the richness of the outer environment, both the world apprehended through the senses and the information about the world stored in long-term memory.

A scientific account of human cognition describes it in terms of several sets of invariants. First, there are the parameters of the inner environment. Then, there are the general control and search-guiding mechanisms that are used over and over again in all task domains. Finally, there are the learning and discovery mechanisms that permit the system to adapt with gradually increasing effectiveness to the particular environment in which it finds itself. The adaptiveness of the human organism, the facility with which it acquires new representations and strategies and becomes adept in dealing with highly specialized environments, makes it an elusive and fascinating target of our scientific inquiries—and the very prototype of the artificial.

5

The Science of Design: Creating the Artificial

Historically and traditionally, it has been the task of the science disciplines to teach about natural things: how they are and how they work. It has been the task of engineering schools to teach about artificial things: how to make artifacts that have desired properties and how to design.

Engineers are not the only professional designers. Everyone designs who devises courses of action aimed at changing existing situations into preferred ones. The intellectual activity that produces material artifacts is no different fundamentally from the one that prescribes remedies for a sick patient or the one that devises a new sales plan for a company or a social welfare policy for a state. Design, so construed, is the core of all professional training; it is the principal mark that distinguishes the professions from the sciences. Schools of engineering, as well as schools of architecture, business, education, law, and medicine, are all centrally concerned with the process of design.

In view of the key role of design in professional activity, it is ironic that in this century the natural sciences almost drove the sciences of the artificial from professional school curricula, a development that peaked about two or three decades after the Second World War. Engineering schools gradually became schools of physics and mathematics; medical schools became schools of biological science; business schools became schools of finite mathematics. The use of adjectives like "applied" concealed, but did not change, the fact. It simply meant that in the professional schools those topics were selected from mathematics and the natural sciences for emphasis which were thought to be most nearly relevant to professional practice. It did not mean that design continued to be taught, as distinguished from analysis.

The movement toward natural science and away from the sciences of the artificial proceeded further and faster in engineering, business, and medicine than in the other professional fields I have mentioned, though it has by no means been absent from schools of law, journalism, and library science. The stronger universities were more deeply affected than the weaker, and the graduate programs more than the undergraduate. During that time few doctoral dissertations in first-rate professional schools dealt with genuine design problems, as distinguished from problems in solid-state physics or stochastic processes. I have to make partial exceptions—for reasons I shall mention—of dissertations in computer science and management science, and there were undoubtedly some others, for example, in chemical engineering.

Such a universal phenomenon must have had a basic cause. It did have a very obvious one. As professional schools, including the independent engineering schools, were more and more absorbed into the general culture of the university, they hankered after academic respectability. In terms of the prevailing norms, academic respectability calls for subject matter that is intellectually tough, analytic, formalizable, and teachable. In the past much, if not most, of what we knew about design and about the artificial sciences was intellectually soft, intuitive, informal, and cookbooky. Why would anyone in a university stoop to teach or learn about designing machines or planning market strategies when he could concern himself with solid-state physics? The answer has been clear: he usually wouldn't.

The damage to professional competence caused by the loss of design from professional curricula gradually gained recognition in engineering and medicine and to a lesser extent in business. Some schools did not think it a problem (and a few still do not), because they regarded schools of applied science as a superior alternative to the trade schools of the past. If that were the choice, we could agree.[1] But neither alternative is

1. That was in fact the choice in our engineering schools a generation ago. The schools needed to be purged of vocationalism; and a genuine science of design did not exist even in a rudimentary form as an alternative. Hence, introducing more fundamental science was the road forward. This was a main theme in Karl Taylor Compton's presidential inaugural address at MIT in 1930:

I hope . . . that increasing attention in the Institute may be given to the fundamental sciences; that they may achieve as never before the spirit and results of re-

satisfactory. The older kind of professional school did not know how to educate for professional design at an intellectual level appropriate to a university; the newer kind of school nearly abdicated responsibility for training in the core professional skill. Thus we were faced with a problem of devising a professional school that could attain two objectives simultaneously: education in both artificial and natural science at a high intellectual level. This too is a problem of design—organizational design.

The kernel of the problem lies in the phrase "artificial science." The previous chapters have shown that a science of artificial phenomena is always in imminent danger of dissolving and vanishing. The peculiar properties of the artifact lie on the thin interface between the natural laws within it and the natural laws without. What can we say about it? What is there to study besides the boundary sciences—those that govern the means and the task environment?

The artificial world is centered precisely on this interface between the inner and outer environments; it is concerned with attaining goals by adapting the former to the latter. The proper study of those who are concerned with the artificial is the way in which that adaptation of means to environments is brought about—and central to that is the process of design itself. The professional schools can reassume their professional responsibilities just to the degree that they discover and teach a science of design, a body of intellectually tough, analytic, partly formalizable, partly empirical, teachable doctrine about the design process.

It is the thesis of this chapter that such a science of design not only is possible but also has been emerging since the mid-1970s. In fact, it is fair to say that the first edition of this book, published in 1969, was influential in its development, serving as a call to action and outlining the form that the action could take. At Carnegie Mellon University, one of the first engineering schools to move toward research on the process of design, the

---

search; that all courses of instruction may be examined carefully to see where training in details has been unduly emphasized at the expense of the more powerful training in all-embracing fundamental principles.

Notice that President Compton's emphasis was on "fundamental," an emphasis as sound today as it was in 1930. What is called for is not a departure from the fundamental but an inclusion in the curriculum of the fundamental in engineering along with the fundamental in natural science. That was not possible in 1930; but it is possible today.

initial step was to form a Design Research Center, about 1975. The Center (since 1985 called the "Engineering Design Research Center") facilitated collaboration among the faculty and students undertaking research on the science and practice of design and developed elements of a theory of design that found their way back into the undergraduate and graduate curricula. The Center continues to play an important role in the modernization and strengthening of education and research in design at Carnegie Mellon and elsewhere in the United States.

In substantial part, design theory is aimed at broadening the capabilities of computers to aid design, drawing upon the tools of artificial intelligence and operations research. Hence, research on many aspects of computer-aided design is being pursued with growing intensity in computer science, engineering and architecture departments, and in operations research groups in business schools. The need to make design theory explicit and precise in order to introduce computers into the process has been the key to establishing its academic acceptability—its appropriateness for a university. In the remainder of this chapter I will take up some of the topics that need to be incorporated in a theory of design and in instruction in design.

The Logic of Design: Fixed Alternatives

We must start with some questions of logic.[2] The natural sciences are concerned with how things are. Ordinary systems of logic—the standard propositional and predicate calculi, say—serve these sciences well. Since the concern of standard logic is with declarative statements, it is well suited for assertions about the world and for inferences from those assertions.

Design, on the other hand, is concerned with how things ought to be, with devising artifacts to attain goals. We might question whether the

2. I have treated the question of logical formalism for design at greater length in two earlier papers: "The Logic of Rational Decision," *British Journal for the Philosophy of Science, 16*(1965):169–186; and "The Logic of Heuristic Decision Making," in Nicholas Rescher (ed.), *The Logic of Decision and Action* (Pittsburgh: University of Pittsburgh Press, 1967), pp. 1–35. The present discussion is based on these two papers, which have been reprinted as chapters 3.1 and 3.2 in my *Models of Discovery* (Dordrecht: D. Reidel Pub. Co., 1977).

forms of reasoning that are appropriate to natural science are suitable also for design. One might well suppose that introduction of the verb "should" may require additional rules of inference, or modification of the rules already imbedded in declarative logic.

Paradoxes of Imperative Logic

Various "paradoxes" have been constructed to demonstrate the need for a distinct logic of imperatives, or a normative, deontic logic. In ordinary logic from "Dogs are pets" and "Cats are pets," one can infer "Dogs and cats are pets." But from "Dogs are pets," "Cats are pets," and "You should keep pets," can one infer "You should keep cats and dogs"? And from "Give me needle and thread!" can one deduce, in analogy with declarative logic, "Give me needle or thread!"? Easily frustrated people would perhaps rather have neither needle nor thread than one without the other, and peace-loving people, neither cats nor dogs, rather than both.

As a response to these challenges of apparent paradox, there have been developed a number of constructions of modal logic for handling "shoulds," "shalts," and "oughts" of various kinds. I think it is fair to say that none of these systems has been sufficiently developed or sufficiently widely applied to demonstrate that it is adequate to handle the logical requirements of the process of design.

Fortunately, such a demonstration is really not essential, for it can be shown that the requirements of design can be met fully by a modest adaptation of ordinary declarative logic. Thus a special logic of imperatives is unnecessary.

I should like to underline the word "unnecessary," which does not mean "impossible." Modal logics can be shown to exist in the same way that giraffes can—namely, by exhibiting some of them. The question is not whether they exist, but whether they are needed for, or even useful for, design.

Reduction to Declarative Logic

The easiest way to discover what kinds of logic are needed for design is to examine what kinds of logic designers use when they are being careful about their reasoning. Now there would be no point in doing this if designers were always sloppy fellows who reasoned loosely, vaguely, and intuitively. Then we might say that whatever logic they used was not the logic they *should* use.

However, there exists a considerable area of design practice where standards of rigor in inference are as high as one could wish. I refer to the domain of so-called "optimization methods," most highly developed in statistical decision theory and management science but acquiring growing importance also in engineering design theory. The theories of probability and utility, and their intersection, have received the painstaking attention not only of practical designers and decision makers but also of a considerable number of the most distinguished logicians and mathematicians of recent generations. F. P. Ramsey, B. de Finetti, A. Wald, J. von Neumann, J. Neyman, K. Arrow, and L. J. Savage are examples.

The logic of optimization methods can be sketched as follows: The "inner environment" of the design problem is represented by a set of given alternatives of action. The alternatives may be given *in extenso:* more commonly they are specified in terms of *command variables* that have defined domains. The "outer environment" is represented by a set of parameters, which may be known with certainty or only in terms of a probability distribution. The goals for adaptation of inner to outer environment are defined by a utility function—a function, usually scalar, of the command variables and environmental parameters—perhaps supplemented by a number of constraints (inequalities, say, between functions of the command variables and environmental parameters). The optimization problem is to find an admissible set of values of the command variables, compatible with the constraints, that maximize the utility function for the given values of the environmental parameters. (In the probabilistic case we might say, "maximize the expected value of the utility function," for instance, instead of "maximize the utility function.")

A stock application of this paradigm is the so-called "diet problem" shown in figure 6. A list of foods is provided, the command variables being quantities of the various foods to be included in the diet. The environmental parameters are the prices and nutritional contents (calories, vitamins, minerals, and so on) of each of the foods. The utility function is the cost (with a minus sign attached) of the diet, subject to the constraints, say, that it not contain more than 2,000 calories per day, that it

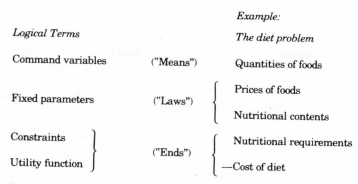

Logical Terms		Example: The diet problem
Command variables	("Means")	Quantities of foods
Fixed parameters	("Laws")	Prices of foods / Nutritional contents
Constraints / Utility function	("Ends")	Nutritional requirements / —Cost of diet

Constraints characterize the inner environment; parameters characterize the outer environment.

Problem: Given the constraints and fixed parameters, find values of the command variables that maximize utility.

Figure 6
The paradigm for imperative logic

meet specified minimum needs for vitamins and minerals, and that rutabaga not be eaten more than once a week. The constraints may be viewed as characterizing the inner environment. The problem is to select the quantities of foods that will meet the nutritional requirements and side conditions at the given prices for the lowest cost.

The diet problem is a simple example of a class of problems that are readily handled, even when the number of variables is exceedingly large, by the mathematical formalism known as linear programming. I shall come back to the technique a little later. My present concern is with the logic of the matter.

Since the optimization problem, once formalized, is a standard mathematical problem—to maximize a function subject to constraints—it is evident that the logic used to deduce the answer is the standard logic of the predicate calculus on which mathematics rests. How does the formalism avoid making use of a special logic of imperatives? It does so by dealing with sets of *possible worlds:* First consider all the possible worlds that meet the constraints of the outer environment; then find the particular world in the set that meets the remaining constraints of the goal and maximizes the utility function. The logic is exactly the same as if we were to adjoin the goal constraints and the maximization requirement, as new "natural laws," to the existing natural laws embodied in the environmental conditions.[3] We simply ask what values the command variables *would* have in a world meeting all these conditions and conclude that these are the values the command variables *should* have.

Computing the Optimum

Our discussion thus far has already provided us with two central topics for the curriculum in the science of design:

1. *Utility theory and statistical decision theory as a logical framework for rational choice among given alternatives.*
2. *The body of techniques for actually deducing which of the available alternatives is the optimum.*

Only in trivial cases is the computation of the optimum alternative an easy matter (Recall Chapter 2). If utility theory is to have application to real-life design problems, it must be accompanied by tools for actually making the computations. The dilemma of the rational chess player is familiar to all. The optimal strategy in chess is easily demonstrated: simply assign a value of $+1$ to a win, 0 to a draw, -1 to a loss; consider all possible courses of play; minimax backward from the outcome of each, assuming each player will take the most favorable move at any given point. This procedure will determine what move to make now. The only trouble is that the computations required are astronomical (the number 10^{120} is often mentioned in this context) and hence cannot be carried out—not by humans, not by existing computers, not by prospective computers.

A theory of design as applied to the game of chess would encompass not only the utopian minimax principle but also some practicable pro-

---

3. The use of the notion of "possible worlds" to embed the logic of imperatives in declarative logic goes back at least to Jørgen Jørgensen, "Imperatives and Logic," *Erkenntnis,* 7(1937-1938):288–296. See also my *Administrative Behavior* (New York: Macmillan, 1947), chapter 3. Typed logics can be used to distinguish, as belonging to different types, statements that are true under different conditions (i.e., in different possible worlds), but, as my example shows, even this device is not usually needed. Each new equation or constraint we introduce into a system reduces the set of possible states to a subset of those previously possible.

cedures for finding good moves in actual board positions in real time, within the computational capacities of real human beings or real computers. The best procedures of this kind that exist today are still those stored in the memories of grandmasters, having the characteristics I described in chapters 3 and 4. But there are now several computer programs that can rather regularly defeat all but a few of the strongest human grandmasters. Even these programs do not possess anything like the chess knowledge of human masters, but succeed by a combination of brute-force computation (sometimes hundreds of millions of variations are analysed) with a good deal of "book" knowledge of opening variations and a reasonably sophisticated criterion function for evaluating positions.

The second topic then for the curriculum in the science of design consists in the efficient computational techniques that are available for actually finding optimum courses of action in real situations, or reasonable approximations to real situations. As I mentioned in chapter 2, that topic has a number of important components today, most of them developed—at least to the level of practical application—within the past years. These include linear programming theory, dynamic programming, geometric programming, queuing theory, and control theory.

Finding Satisfactory Actions

The subject of computational techniques need not be limited to optimization. Traditional engineering design methods make much more use of inequalities—specifications of satisfactory performance—than of maxima and minima. So-called "figures of merit" permit comparison between designs in terms of "better" and "worse" but seldom provide a judgment of "best." For example, I may cite the root-locus methods employed in the design of control systems.

Since there did not seem to be any word in English for decision methods that look for good or satisfactory solutions instead of optimal ones, some years ago I introduced the term "satisficing" to refer to such procedures. Now no one in his right mind will satisfice if he can equally well optimize; no one will settle for good or better if he can have best. But that is not the way the problem usually poses itself in actual design situations.

In chapter 2 I argued that in the real world we usually do not have a choice between satisfactory and optimal solutions, for we only rarely have a method of finding the optimum. Consider, for example, the well-known combinatorial problem called the traveling salesman problem: given the geographical locations of a set of cities, find the routing that will take a salesman to all the cities with the shortest mileage.[4] For this problem there is a straightforward optimizing algorithm (analogous to the minimax algorithm for chess): try all possible routings, and pick the shortest. But for any considerable number of cities, the algorithm is computationally infeasible (the number of routes through N cities will be $N!$). Although some ways have been found for cutting down the length of the search, no algorithm has been discovered sufficiently powerful to solve the traveling salesman problem with a tolerable amount of computing for a set of, say, fifty cities.

Rather than keep our salesman at home, we shall prefer of course to find a satisfactory, if not optimal, routing for him. Under most circumstances, common sense will probably arrive at a fairly good route, but an even better one can often be found by one or another of several heuristic methods.

An earmark of all these situations where we satisfice for inability to optimize is that, although the set of available alternatives is "given" in a certain abstract sense (we can define a generator guaranteed to generate all of them eventually), it is not "given" in the only sense that is practically relevant. We cannot within practicable computational limits generate all the admissible alternatives and compare their respective merits. Nor can we recognize the best alternative, even if we are fortunate enough to generate it early, until we have seen all of them. We satisfice by looking for alternatives in such a way that we can generally find an acceptable one after only moderate search.

Now in many satisficing situations, the expected length of search for an alternative meeting specified standards of acceptability depends on how high the standards are set, but it depends hardly at all on the total size of the universe to be searched. The time required for a search through a haystack for a needle sharp enough to sew with depends on the density of distribution of sharp needles but not on the total size of the stack.

4. "The traveling salesman problem" and a number of closely analogous combinatorial problems—such as the "warehouse location problem"—have considerable practical importance, for instance, in siting central power stations for an interconnected grid.

Hence, when we use satisficing methods, it often does not matter whether or not the total set of admissible alternatives is "given" by a formal but impracticable algorithm. It often does not even matter how big that set is. For this reason satisficing methods may be extendable to design problems in that broad range where the set of alternatives is not "given" even in the quixotic sense that it is "given" for the traveling salesman problem. Our next task is to examine this possibility.

The Logic of Design: Finding Alternatives

When we take up the case where the design alternatives are not given in any constructive sense but must be synthesized, we must ask once more whether any new forms of reasoning are involved in the synthesis, or whether again the standard logic of declarative statements is all we need.

In the case of optimization we asked: "Of all possible worlds (those attainable for some admissible values of the action variables), which is the best (yields the highest value of the criterion function)?" As we saw, this is a purely empirical question, calling only for facts and ordinary declarative reasoning to answer it.

In this case, where we are seeking a satisfactory alternative, once we have found a candidate we can ask: "Does this alternative satisfy all the design criteria?" Clearly this is also a factual question and raises no new issues of logic. But how about the process of *searching* for candidates? What kind of logic is needed for the search?

Means-Ends Analysis

The condition of any goal-seeking system is that it is connected to the outside environment through two kinds of channels: the afferent, or sensory, channels through which it receives information about the environment and the efferent, or motor, channels through which it acts on the environment.[5] The system must have some means of storing in its memory information about states of the world—afferent, or sensory, information—

5. Notice that we are not saying that the two kinds of channels operate independently of each other, since they surely do not in living organisms, but that we can distinguish conceptually, and to some extent neurologically, between the incoming and outgoing flows.

and information about actions—efferent, or motor, information. Ability to attain goals depends on building up associations, which may be simple or very complex, between particular changes in states of the world and particular actions that will (reliably or not) bring these changes about. In chapter 4 we described these associations as productions.

Except for a few built-in reflexes, an infant has no basis for correlating its sensory information with its actions. A very important part of its early learning is that particular actions or sequences of actions will bring about particular changes in the state of the sensed world. Until the infant builds up this knowledge, the world of sense and the motor world are two entirely separate, entirely unrelated worlds. Only as it begins to acquire experience as to how elements of the one relate to elements of the other can it act purposefully on the world.

The computer problem-solving program called GPS, designed to model some of the main features of human problem solving, exhibits in stark form how goal-directed action depends on building this kind of bridge between the afferent and the efferent worlds. On the afferent, or sensory, side, GPS must be able to represent desired situations or desired objects as well as the present situation. It must be able also to represent *differences* between the desired and the present. On the efferent side, GPS must be able to represent *actions* that change objects or situations. To behave purposefully, GPS must be able to select from time to time those particular actions that are likely to remove the particular differences between desired and present states that the system detects. In the machinery of GPS, this selection is achieved through a *table of connections,* which associates with each kind of detectable difference those actions that are relevant to reducing that difference. These are its associations, in the form of productions, which relate the afferent to the efferent world. Since reaching a goal generally requires a sequence of actions, and since some attempts may be ineffective, GPS must also have means for detecting the progress it is making (the changes in the differences between the actual and the desired) and for trying alternate paths.

The Logic of Search
GPS then is a system that searches selectively through a (possibly large) environment in order to discover and assemble sequences of actions that

will lead it from a given situation to a desired situation. What are the rules of logic that govern such a search? Is anything more than standard logic involved? Do we require a modal logic to rationalize the process?

Standard logic would seem to suffice. To represent the relation between the afferent and the efferent worlds, we conceive GPS as moving through a large maze. The nodes of the maze represent situations, described afferently; the paths joining one node to another are the actions, described as motor sequences, that will transform the one situation into the other. At any given moment GPS is always faced with a single question: "What action shall I try next?" Since GPS has some imperfect knowledge about the relations of actions to changes in the situation, this becomes a question of choice under uncertainty of a kind already discussed in a previous section.

It is characteristic of the search for alternatives that the solution, the complete action that constitutes the final design, is built from a sequence of component actions. The enormous size of the space of alternatives arises out of the innumerable ways in which the component actions, which need not be very numerous, can be combined into sequences.

Much is gained by considering the component actions in place of the sequences that constitute complete actions, because the situation when viewed afferently usually factors into components that match at least approximately the component actions derived from an efferent factorization. The reasoning implicit in GPS is that, if a desired situation differs from a present situation by differences $D_1, D_2, \ldots, D_n$, and if action A_1 removes differences of type D_1, action A_2 removes differences of type D_2, and so on, then the present situation can be transformed into the desired situation by performing the sequence of actions $A_1 A_2 \ldots A_n$.

This reasoning is by no means valid in terms of the rules of standard logic in all possible worlds. Its validity requires some rather strong assumptions about the independence of the effects of the several actions on the several differences. One might say that the reasoning is valid in worlds that are "additive" or "factorable" in a certain sense. (The air of paradox about the cat-dog and needle-thread examples cited earlier arises precisely from the nonadditivity of the actions in these two cases. The first is, in economists' language, a case of decreasing returns; the second, a case of increasing returns.)

Now the real worlds to which problem solvers and designers address themselves are seldom completely additive in this sense. Actions have side consequences (may create new differences) and sometimes can only be taken when certain side conditions are satisfied (call for removal of other differences before they become applicable). Under these circumstances one can never be certain that a partial sequence of actions that accomplishes *certain* goals can be augmented to provide a solution that satisfies *all* the conditions and attains *all* the goals (even though they be satisficing goals) of the problem.

For this reason problem-solving systems and design procedures in the real world do not merely *assemble* problem solutions from components but must *search* for appropriate assemblies. In carrying out such a search, it is often efficient to divide one's eggs among a number of baskets—that is, not to follow out one line until it succeeds completely or fails definitely but to begin to explore several tentative paths, continuing to pursue a few that look most promising at a given moment. If one of the active paths begins to look less promising, it may be replaced by another that had previously been assigned a lower priority.

Our discussion of design when the alternatives are not given has yielded at least three additional topics for instruction in the science of design:

3. *Adaptation of standard logic to the search for alternatives.* Design solutions are sequences of actions that lead to possible worlds satisfying specified constraints. With satisficing goals the sought-for possible worlds are seldom unique; the search is for *sufficient,* not *necessary,* actions for attaining goals.

4. *The exploitation of parallel, or near-parallel, factorizations of differences.* Means-end analysis is an example of a broadly applicable problem-solving technique that exploits this factorization.

5. *The allocation of resources for search to alternative, partly explored action sequences.* I should like to elaborate somewhat on this last-mentioned topic.

Design as Resource Allocation

There are two ways in which design processes are concerned with the allocation of resources. First, conservation of scarce resources may be one of the criteria for a satisfactory design. Second, the design process itself involves management of the resources of the designer, so that his efforts will not be dissipated unnecessarily in following lines of inquiry that prove fruitless.

There is nothing special that needs to be said here about resource conservation—cost minimization, for example, as a design criterion. Cost minimization has always been an implicit consideration in the design of engineering structures, but until a few years ago it generally *was* only implicit, rather than explicit. More and more cost calculations have been brought explicitly into the design procedure, and a strong case can be made today for training design engineers in that body of technique and theory that economists know as "cost-benefit analysis."

An Example from Highway Design

The notion that the costs of designing must themselves be considered in guiding the design process began to take root only as formal design procedures have developed, and it still is not universally applied. An early example, but still a very good one, of incorporating design costs in the design process is the procedure, developed by Marvin L. Manheim as a doctoral thesis at MIT, for solving highway location problems.[6]

Manheim's procedure incorporates two main notions: first, the idea of specifying a design progressively from the level of very general plans down to determining the actual construction; second, the idea of attaching values to plans at the higher levels as a basis for deciding which plans to pursue to levels of greater specificity.

In the case of highway design the higher-level search is directed toward discovering "bands of interest" within which the prospects of finding a good specific route are promising. Within each band of interest one or more locations is selected for closer examination. Specific designs are then developed for particular locations. The scheme is not limited of course to this specific three-level division, but it can be generalized as appropriate.

Manheim's scheme for deciding which alternatives to pursue from one level to the next is based on assigning costs to each of the design activities and estimating highway costs for each of the higher-level plans. The

6. Marvin L. Manheim, *Hierarchical Structure: A Model of Design and Planning Processes* (Cambridge: The MIT Press, 1966).

highway cost associated with a plan is a prediction of what the cost would be for the actual route if that plan were particularized through subsequent design activity. In other words, it is a measure of how "promising" a plan is. Those plans are then pursued to completion that look most promising after the prospective design costs have been offset against them.

In the particular method that Manheim describes, the "promise" of a plan is represented by a probability distribution of outcomes that would ensue if it were pursued to completion. The distribution must be estimated by the engineer—a serious weakness of the method—but, once estimated, it can be used within the framework of Bayesian decision theory. The particular probability model used is not the important thing about the method; other methods of valuation without the Bayesian superstructure might be just as satisfactory.

In the highway location procedure the evaluation of higher-level plans performs two functions. First, it answers the question, "Where shall I search next?" Second, it answers the question, "When shall I stop the search and accept a solution as satisfactory?" Thus it is both a steering mechanism for the search and a satisficing criterion for terminating the search.

Schemes for Guiding Search

Let us generalize the notion of schemes for guiding search activity beyond Manheim's specific application to a highway location problem and beyond his specific guidance scheme based on Bayesian decision theory. Consider the typical structure of a problem-solving program. The program begins to search along possible paths, storing in memory a "tree" of the paths it has explored. Attached to the end of each branch—each partial path—is a number that is supposed to express the "value" of that path.

But the term "value" is really a misnomer. A partial path is not a solution of the problem, and a path has a "true" value of zero unless it leads toward a solution. Hence it is more useful to think of the values as estimates of the gain to be expected from further search along the path than to think of them as "values" in any more direct sense. For example, it may be desirable to attach a relatively high value to a partial exploration that *may* lead to a very good solution but with a low probability. If the prospect fades on further exploration, only the cost of the search has been lost. The disappointing outcome need not be accepted, but an alternative path may be taken instead. Thus the scheme for attaching values to partial paths may be quite different from the evaluation function for proposed complete solutions.[7]

When we recognize that the purpose of assigning values to incomplete paths is to guide the choice of the next point for exploration, it is natural to generalize even further. All kinds of information gathered in the course of search may be of value in selecting the next step in search. We need not limit ourselves to valuations of partial search paths.

For example, in a chess-playing program an exploration may generate a continuation move different from any that was proposed by the initial move generator. Whatever the context—the branch of the search tree— on which the move was actually generated, it can now be removed from the context and considered in the context of other move sequences. Such a scheme was added on a limited basis by Baylor to MATER, a program for discovering check-mating combinations in chess, and it proved to enhance the program's power significantly.[8]

Thus search processes may be viewed—as they have been in most discussions of problem solving—as processes for seeking a problem solution. But they can be viewed more generally as processes for gathering information about problem structure that will ultimately be valuable in discovering a problem solution. The latter viewpoint is more general than the former in a significant sense, in that it suggests that information obtained along any particular branch of a search tree may be used in many contexts besides the one in which it was generated. Only a few problem-solving programs exist today that can be regarded as moving even a modest distance from the earlier, more limited viewpoint to the newer one.[9]

7. That this point is not obvious can be seen from the fact that most chess-playing programs have used similar or identical evaluation procedures both to guide search and to evaluate the positions reached at the ends of paths.

8. George W. Baylor and Herbert A. Simon, "A Chess Mating Combinations Program," *Proceedings of the Spring Joint Computer Conference,* Boston, April 26–28, (1966):431–447 (Washington: Spartan Books, 1966), reprinted in *Models of Thought,* chapter 4.3.

9. A formal theory of the optimal choice of search paths can be found in H. A. Simon and J. B. Kadane, "Optimal Problem-Solving Search: All-or-none Solutions," *Artificial Intelligence,* 6(1975):235–247.

The Shape of the Design: Hierarchy

In my first chapter I gave some reasons why complex systems might be expected to be constructed in a hierarchy of levels, or in a boxes-within-boxes form. The basic idea is that the several components in any complex system will perform particular subfunctions that contribute to the overall function. Just as the "inner environment" of the whole system may be defined by describing its functions, without detailed specification of its mechanisms, so the "inner environment" of each of the subsystems may be defined by describing the functions of that subsystem, without detailed specification of *its* submechanisms.[10]

To design such a complex structure, one powerful technique is to discover viable ways of decomposing it into semi-independent components corresponding to its many functional parts. The design of each component can then be carried out with some degree of independence of the design of others, since each will affect the others largely through its function and independently of the details of the mechanisms that accomplish the function.[11]

There is no reason to expect that the decomposition of the complete design into functional components will be unique. In important instances there may exist alternative feasible decompositions of radically different kinds. This possibility is well known to designers of administrative organizations, where work can be divided up by subfunctions, by subprocesses, by subareas, and in other ways. Much of classical organization theory in fact was concerned precisely with this issue of alternative decompositions of a collection of interrelated tasks.

The Generator-Test Cycle

One way of considering the decomposition, but acknowledging that the interrelations among the components cannot be ignored completely, is to think of the design process as involving, first, the generation of alterna-

---

10. I have developed this argument at greater length in my essay "The Architecture of Complexity," chapter 8.

11. For a recent discussion of functional analysis in design, see Clive L. Dym, *Engineering Design* (New York, NY: Cambridge University Press, 1994), pp. 134–139.

tives and, then, the testing of these alternatives against a whole array of requirements and constraints. There need not be merely a single generate-test cycle, but there can be a whole nested series of such cycles. The generators implicitly define the decomposition of the design problem, and the tests guarantee that important indirect consequences will be noticed and weighed. Alternative decompositions correspond to different ways of dividing the responsibilities for the final design between generators and tests.

To take a greatly oversimplified example, a series of generators may generate one or more possible outlines and schemes of fenestration for a building, while tests may be applied to determine whether needs for particular kinds of rooms can be met within the outlines generated. Alternatively the generators may be used to evolve the structure of rooms, while tests are applied to see whether they are consistent with an acceptable over-all shape and design. The house can be designed from the outside in or from the inside out.[12]

Alternatives are also open, in organizing the design process, as to how far development of possible subsystems will be carried before the over-all coordinating design is developed in detail, or vice-versa, how far the over-all design should be carried before various components, or possible components, are developed. These alternatives of design are familiar to architects. They are familiar also to composers, who must decide how far the architectonics of a musical structure will be evolved before some of the component musical themes and other elements have been invented. Computer programmers face the same choices, between working downward from executive routines to subroutines or upward from component subroutines to a coordinating executive.

A theory of design will include principles for deciding such questions of precedence and sequence in the design process. As examples, the approach to designing computer programs called structured programming is concerned in considerable part with attending to design subproblems

12. I am indebted to John Grason for many ideas on the topic of this section. J. Grason, "Fundamental Description of a Floor Plan Design Program," EDRA1, *Proceedings of the First Environmental Design Association Conference,* H. Sanoff and S. Cohn (eds.), North Carolina State University, 1970.

in the proper order (usually top-down); and much instruction in schools of architecture focuses on the same concerns.

Process as a Determinant of Style

When we recall that the process will generally be concerned with finding a satisfactory design, rather than an optimum design, we see that sequence and the division of labor between generators and tests can affect not only the efficiency with which resources for designing are used but also the nature of the final design as well. What we ordinarily call "style" may stem just as much from these decisions about the design process as from alternative emphases on the goals to be realized through the final design.[13] An architect who designs buildings from the outside in will arrive at quite different buildings from one who designs from the inside out, even though both of them might agree on the characteristics that a satisfactory building should possess.

When we come to the design of systems as complex as cities, or buildings, or economies, we must give up the aim of creating systems that will optimize some hypothesized utility function, and we must consider whether differences in style of the sort I have just been describing do not represent highly desirable variants in the design process rather than alternatives to be evaluated as "better" or "worse." Variety, within the limits of satisfactory constraints, may be a desirable end in itself, among other reasons, because it permits us to attach value to the search as well as its outcome—to regard the design process as itself a valued activity for those who participate in it.

We have usually thought of city planning as a means whereby the planner's creative activity could build a system that would satisfy the needs of a populace. Perhaps we should think of city planning as a valuable creative activity in which many members of a community can have the opportunity of participating—if we have wits to organize the process that way. I shall have more to say on these topics in the next chapter.

However that may be, I hope I have illustrated sufficiently that both the shape of the design and the shape and organization of the design process

13. H. A. Simon, "Style in Design," *Proceedings of the 2nd Annual Conference of the Environmental Design Research Association,* Pittsburgh, PA: Carnegie Mellon University (1971), pp. 1–10.

are essential components of a theory of design. These topics constitute the sixth item in my proposed curriculum in design:

6. *The organization of complex structures and its implication for the organization of design processes.*

Representation of the Design

I have by no means surveyed all facets of the emerging science of design. In particular I have said little about the influence of problem representation on design. Although the importance of the question is recognized today, we are still far from a systematic theory of the subject—in particular, a theory that would tell us how to generate effective problem representations.[14] I shall cite one example, to make clear what I mean by "representation."

Here are the rules of a game, which I shall call number scrabble. The game is played by two people with nine cards—let us say the ace through the nine of hearts. The cards are placed in a row, face up, between the two players. The players draw alternately, one at a time, selecting any one of the cards that remain in the center. The aim of the game is for a player to make up a "book," that is, a set of exactly three cards whose spots add to 15, before his opponent can do so. The first player who makes a book wins; if all nine cards have been drawn without either player making a book, the game is a draw.

What is a good strategy in this game? How would you go about finding one? If the reader has not already discovered it for himself, let me show how a change in representation will make it easy to play the game well. The magic square here, which I introduced in the third chapter, is made up of the numerals from 1 through 9.

```
4  9  2
3  5  7
8  1  6
```

14. As examples of current thinking about representation see chapters 5 ("Representing Designed Artifacts") and 6 ("Representing Design Processes") in C. L. Dym, *op. cit.,* and chapter 6 ("Representation in Design") in Ömer Akin, *op. cit.* For a more general theoretical discussion, see R. E. Korf, "Toward a Model of Representational Changes," *Artificial Intelligence,* 14(1980):41–78.

Each row, column, or diagonal adds to 15, and every triple of these numerals that add to 15 is a row, column, or diagonal of the magic square. From this, it is obvious that "making a book" in number scrabble is equivalent to getting "three in a row" in the game of tic-tac-toe. But most people know how to play tic-tac-toe well, hence can simply transfer their usual strategy to number scrabble.[15]

Problem Solving as Change in Representation

That representation makes a difference is a long-familiar point. We all believe that arithmetic has become easier since Arabic numerals and place notation replaced Roman numerals, although I know of no theoretic treatment that explains why.

That representation makes a difference is evident for a different reason. All mathematics exhibits in its conclusions only what is already implicit in its premises, as I mentioned in a previous chapter. Hence all mathematical derivation can be viewed simply as change in representation, making evident what was previously true but obscure.

This view can be extended to all of problem solving—solving a problem simply means representing it so as to make the solution transparent.[16] If the problem solving could actually be organized in these terms, the issue of representation would indeed become central. But even if it cannot— if this is too exaggerated a view—a deeper understanding of how representations are created and how they contribute to the solution of problems will become an essential component in the future theory of design.

Spatial Representation

Since much of design, particularly architectural and engineering design, is concerned with objects or arrangements in real Euclidean two-

15. Number scrabble is not the only isomorph of tic-tac-toe. John A. Michon has described another, JAM, which is the dual of tic-tac-toe in the sense of projective geometry. That is, the rows, columns, and diagonals of tic-tac-toe become points in JAM, and the squares of the former become line segments joining the points. The game is won by "jamming" all the segments through a point—a move consists of seizing or jamming a single segment. Other isomorphs of tic-tac-toe are known as well.

16. Saul Amarel, "On the Mechanization of Creative Processes," IEEE *Spectrum* 3(April 1966):112–114.

dimensional or three-dimensional space, the representation of space and of things in space will necessarily be a central topic in a science of design. From our previous discussion of visual perception, it should be clear that "space" inside the head of the designer or the memory of a computer may have very different properties from a picture on paper or a three-dimensional model.

These representational issues have already attracted the attention of those concerned with computer-aided design—the cooperation of human and computer in the design process. As a single example, I may mention Ivan Sutherland's pioneering SKETCHPAD program which allowed geometric shapes to be represented and conditions to be placed on these shapes in terms of constraints, to which they then conformed.[17]

Geometric considerations are also prominent in the attempts to automate completely the design, say, of printed or etched circuits, or of buildings. Grason, for example, in a system for designing house floor plans, constructs an internal representation of the layout that helps one decide whether a proposed set of connections among rooms, selected to meet design criteria for communication, and so on, can be realized in a plane.[18]

The Taxonomy of Representation
An early step toward understanding any set of phenomena is to learn what kinds of things there are in the set—to develop a taxonomy. This step has not yet been taken with respect to representations. We have only a sketchy and incomplete knowledge of the different ways in which problems can be represented and much less knowledge of the significance of the differences.

In a completely pragmatic vein we know that problems can be described verbally, in natural language. They often can be described mathematically, using standard formalisms of algebra, geometry, set theory, analysis, or topology. If the problems relate to physical objects, they (or their solutions) can be represented by floor plans, engineering drawings,

17. I. E. Sutherland, "SKETCHPAD, A Man-Machine Graphical Communication System," *Proceedings, AFIPS Spring Joint Computer Conference, 1963* (Baltimore: Spartan Books), pp. 329–346.

18. See also C. E. Pfefferkorn, "The Design Problem Solver: A System for Designing Equipment or Furniture Layouts," in C. M. Eastman (ed.), *Spatial Synthesis in Computer-Aided Building Design* (London: Applied Science Publishers, 1975).

renderings, or three-dimensional models. Problems that have to do with actions can be attacked with flow charts and programs.

Other items most likely will need to be added to the list, and there may exist more fundamental and significant ways of classifying its members. But even though our classification is incomplete, we are beginning to build a theory of the properties of these representations. The growing theories of computer architectures and programming languages—for example; the work on functional languages and object-oriented languages—illustrate some of the directions that a theory of representations can take. There has also been closely parallel progress, some of it reviewed in chapters 3 and 4, toward understanding the human use of representations in thinking. These topics begin to provide substance for the final subject in our program on the theory of design:

7. *Alternative representations for design problems.*

Summary—Topics in The Theory of Design

My main goal in this chapter has been to show that there already exist today a number of components of a theory of design and a substantial body of knowledge, theoretical and empirical, relating to each. As we draw up our curriculum in design—in the science of the artificial—to take its place by the side of natural science in the whole engineering curriculum, it includes at least the following topics:

THE EVALUATION OF DESIGNS
  1. Theory of evaluation: utility theory, statistical decision theory
  2. Computational methods:
     a. Algorithms for choosing *optimal* alternatives such as linear programming computations, control theory, dynamic programming
     b. Algorithms and heuristics for choosing *satisfactory* alternatives
  3. THE FORMAL LOGIC OF DESIGN: imperative and declarative logics
THE SEARCH FOR ALTERNATIVES
  4. Heuristic search: factorization and means-ends analysis
  5. Allocation of resources for search
  6. THEORY OF STRUCTURE AND DESIGN ORGANIZATION: hierarchic systems
  7. REPRESENTATION OF DESIGN PROBLEMS

In small segments of the curriculum—the theory of evaluation, for example, and the formal logic of design—it is already possible to organize the instruction within a framework of systematic, formal theory. In many other segments the treatment would be more pragmatic, more empirical.

But nowhere do we need to return or retreat to the methods of the cookbook that originally put design into disrepute and drove it from the engineering curriculum. For there exist today a considerable number of examples of actual design processes, of many different kinds, that have been defined fully and cast in the metal, so to speak, in the form of running computer programs: optimizing algorithms, search procedures, and special-purpose programs for designing motors, balancing assembly lines, selecting investment portfolios, locating warehouses, designing highways, diagnosing and treating diseases, and so forth.[19]

Because these computer programs describe complex design processes in complete, painstaking detail, they are open to full inspection and analysis, or to trial by simulation. They constitute a body of empirical phenomena to which the student of design can address himself and which he can seek to understand. There is no question, since these programs exist, of the design process hiding behind the cloak of "judgment" or "experience." Whatever judgment or experience was used in creating the programs must now be incorporated in them and hence be observable. The programs are the tangible record of the variety of schemes that man has devised to explore his complex outer environment and to discover in that environment the paths to his goals.

Role of Design in the Life of the Mind

I have called my topic "the theory of design" and my curriculum a "program in design." I have emphasized its role as complement to the natural

19. A number of these programs are described in Dym, *op. cit.*, and others are discussed in a forthcoming book on *Engineering Design in the Large*, written by faculty associated with the Engineering Design Research Center at Carnegie Mellon University. Dym concludes each chapter of his book with a commentary on other relevant publications. Dym's book has a bibliography of more than 200 items, a majority of them referring to specific design projects and systems; its extent gives some indication of the rate at which the science of design is now progressing.

science curriculum in the total training of a professional engineer—or of any professional whose task is to solve problems, to choose, to synthesize, to decide.

But there is another way in which the theory of design may be viewed in relation to other knowledge. My third and fourth chapters were chapters on psychology—specifically on man's relation to his biological inner environment. The present chapter may also be construed as a chapter on psychology: on man's relation to the complex outer environment in which he seeks to survive and achieve.

All three chapters, so construed, have import that goes beyond the professional work of the person we have called the "designer." Many of us have been unhappy about the fragmentation of our society into two cultures. Some of us even think there are not just two cultures but a large number of cultures. If we regret that fragmentation, then we must look for a common core of knowledge that can be shared by the members of all cultures—a core that includes more significant topics than the weather, sports, automobiles, the care and feeding of children, or perhaps even politics. A common understanding of our relation to the inner and outer environments that define the space in which we live and choose can provide at least part of that significant core.

This may seem an extravagant claim. Let me use the realm of music to illustrate what I mean. Music is one of the most ancient of the sciences of the artificial, and was so recognized by the Greeks. Anything I have said about the artificial would apply as well to music, its composition or its enjoyment, as to the engineering topics I have used for most of my illustrations.

Music involves a formal pattern. It has few (but important) contacts with the inner environment; that is, it is capable of evoking strong emotions, its patterns are detectable by human listeners, and some of its harmonic relations can be given physical and physiological interpretations (though the aesthetic import of these is debatable). As for the outer environment, when we view composition as a problem in design, we encounter just the same tasks of evaluation, of search for alternatives, and of representation that we do in any other design problem. If it pleases us, we can even apply to music some of the same techniques of automatic design by computer that have been used in other fields of design. If computer-composed music has not yet reached notable heights of aesthetic excellence, it deserves, and has already received, serious attention from professional composers and analysts, who do not find it written in tongues alien to them.[20]

Undoubtedly there are tone-deaf engineers, just as there are mathematically ignorant composers. Few engineers and composers, whether deaf, ignorant, or not, can carry on a mutually rewarding conversation about the content of each other's professional work. What I am suggesting is that they *can* carry on such a conversation about design, can begin to perceive the common creative activity in which they are both engaged, can begin to share their experiences of the creative, professional design process.

Those of us who have lived close to the development of the modern computer through gestation and infancy have been drawn from a wide variety of professional fields, music being one of them. We have noticed the growing communication among intellectual disciplines that takes place around the computer. We have welcomed it, because it has brought us into contact with new worlds of knowledge—has helped us combat our own multiple-cultures isolation. This breakdown of old disciplinary boundaries has been much commented upon, and its connection with computers and the information sciences often noted.

But surely the computer, as a piece of hardware, or even as a piece of programmed software, has nothing to do directly with the matter. I have already suggested a different explanation. The ability to communicate across fields—the common ground—comes from the fact that all who use computers in complex ways are using computers to design or to participate in the process of design. Consequently we as designers, or as designers of design processes, have had to be explicit as never before about what is involved in creating a design and what takes place while the creation is going on.

The real subjects of the new intellectual free trade among the many cultures are our own thought processes, our processes of judging, deciding,

20. L. A. Hillier and L. M. Isaacson's *Experimental Music* (New York: McGraw-Hill, 1959), reporting experiments begun more than four decades ago, still provides a good introduction to the subject of musical composition, viewed as design. See also Walter R. Reitman, *Cognition and Thought* (New York: Wiley, 1965), chapter 6, "Creative Problem Solving: Notes from the Autobiography of a Fugue."

choosing, and creating. We are importing and exporting from one intellectual discipline to another ideas about how a serially organized information-processing system like a human being—or a computer, or a complex of men and women and computers in organized cooperation—solves problems and achieves goals in outer environments of great complexity.

The proper study of mankind has been said to be man. But I have argued that people—or at least their intellective component—may be relatively simple, that most of the complexity of their behavior may be drawn from their environment, from their search for good designs. If I have made my case, then we can conclude that, in large part, the proper study of mankind is the science of design, not only as the professional component of a technical education but as a core discipline for every liberally educated person.

6

Social Planning: Designing the Evolving Artifact

In chapter 5 I surveyed some of the modern tools of design that are used by planners and artificers. Even before most of these tools were available to them, ambitious planners often took whole societies and their environments as systems to be refashioned. Some recorded their utopias in books—Plato, Sir Thomas More, Marx. Others sought to realize their plans by social revolution in America, France, Russia, China. Many or most of the large-scale designs have centered on political and economic arrangements, but others have focused on the physical environment—river development plans, for example, reaching from ancient Egypt to the Tennessee Valley to the Indus and back to today's Nile.

As we look back on such design efforts and their implementation, and as we contemplate the tasks of design that are posed in the world today, our feelings are very mixed. We are energized by the great power our technological knowledge bestows on us. We are intimidated by the magnitude of the problems it creates or alerts us to. We are sobered by the very limited success—and sometimes disastrous failure—of past efforts to design on the scale of whole societies. We ask, "If we can go to the Moon, why can't we . . . ?"—not expecting an answer, for we know that going to the Moon was a simple task indeed, compared with some others we have set for ourselves, such as creating a humane society or a peaceful world. Wherein lies the difference?

Going to the Moon was a complex matter along only one dimension: it challenged our technological capabilities. Though it was no mean accomplishment, it was achieved in an exceedingly cooperative environment, employing a single new organization, NASA, that was charged with a single, highly operational goal. With enormous resources provided to it, and operating through well-developed market mechanisms, that organization could draw on the production capabilities and technological sophistication of our whole society. Although several potential side effects of the activity (notably its international political and military significance, and the possibility of technological spinoffs) played a major role in motivating the project, they did not have to enter much into the thoughts of the planners once the goal of placing human beings on the Moon had been set. Moreover these by-product benefits and costs are not what we mean when we say the project was a success. It was a success because people walked on the surface of the Moon. Nor did anyone anticipate what turned out to be one of the more important consequences of these voyages: the vivid new perspective we gained of our place in the universe when we first viewed our own pale, fragile planet from space.

Consider now a quite different example of human design. Some twenty years ago we celebrated the 200th birthday of our nation, and about a decade ago we celebrated the 200th anniversary of the framing of its political constitution. Almost all of us in the free world regard that document as an impressive example of success in human planning. We regard it as a success because the Constitution, much modified and much interpreted, still survives as the framework for our political institutions, and because the society that operates within its framework has provided most of us with a broad range of freedoms and a high level of material comfort.

Both achievements—the voyages to the Moon and the survival of the American Constitution—are triumphs of bounded rationality. A necessary, though not a sufficient, condition of their success was that they were evaluated against limited objectives. I have already argued that point with respect to NASA. As to the founding fathers it is instructive to examine their own views of their goals, reflected in *The Federalist* and the surviving records of the constitutional convention.[1] What is striking about these documents is their practical sense and the awareness they exude of the limits of foresight about large human affairs. Most of the framers of the Constitution accepted very restricted objectives for their artifact—princi-

---

1. The authors of *The Federalist* were Madison, Hamilton, and Jay, but principally the first named. My edition is that edited by P. L. Ford (New York: Holt, 1898). Madison's notes are our chief source on the proceedings of the convention.

pally the preservation of freedom in an orderly society. Moreover they did not postulate a new man to be produced by the new institutions but accepted as one of their design constraints the psychological characteristics of men and women as they knew them, their selfishness as well as their common sense. In their own cautious words (*The Federalist*, no. 55), "As there is a degree of depravity in mankind which requires a certain degree of circumspection and distrust, so there are other qualities in human nature which justify a certain portion of esteem and confidence."

These examples illustrate some of the characteristics and complexities of designing artifacts on a societal scale. The success of planning on such a scale may call for modesty and restraint in setting the design objectives and drastic simplification of the real-world situation in representing it for purposes of the design process. Even with restraint and simplification difficult obstacles must usually be surmounted to reach the design objectives. The obstacles and some of the techniques for overcoming them provide the main subject of this chapter.

Our first topic will be problem representation; our second, ways of accommodating to the inadequacies that can be expected in data; our third, how the nature of the client affects planning; our fourth, limits on the planner's time and attention; and our fifth, the ambiguity and conflict of goals in societal planning. These topics, which can be viewed as a budget of obstacles or alternatively as a budget of planning requirements, will suggest to us some additions to the curriculum in design outlined in the last chapter.

Representing the Design Problem

In the previous chapter representation was discussed mainly in the context of relatively well-structured, middle-sized tasks. Representation problems take on new dimensions where social design is involved.

Organization as Representation

In 1948 the U.S. government took a bold initiative to restore the postwar economies of the nations of western Europe, the so-called "Marshall Plan," which was implemented through the Economic Cooperation

Administration (ECA).[2] An initial task in carrying out the plan was to shape the ECA organization. The answer to the organizational question depended on how one conceptualized the program. At least six different, and largely contradictory, conceptions were offered for the agency by the persons who were initially recruited to organize and manage it.

Congress had appropriated $5.3 billion for the first year's operations. Some thought that the task was to screen shopping lists proposed by the European nations to make sure the lists contained what was really "needed" (commodity screening approach). Others thought the task was to determine the "dollar gap" in each nation's balance of payments and to authorize funds to close that gap (balance of trade approach). Others thought that the main task was to build up a strong deliberative institution in Europe, so that the recipient nations could make their own plans for use of the funds and thereby strengthen their collaboration (European cooperation approach). Others thought that decisions should be made primarily through bilateral agreements between the United States and each of the recipient nations (bilateral pledge approach). Others thought that at least the portion of the appropriation that was earmarked for loans ($1 billion) should be handled on a project basis, each project being evaluated for its soundness as an investment (investment bank approach). Others thought that the ECA should have a policy organ for making broad decisions, then a number of administrative organs for implementing them (policy and administration approach). Each of these representations had some basis in the congressional legislation establishing the ECA.

With a little reflection it is easy to see that very different assistance plans would result from implementing these different approaches, with very different economic and political consequences for the European nations and the United States. Conceptualizing the problem in a particular way implied organizing the agency in a manner consistent with that conceptualization. And different organizations would lead inevitably to the implementation of quite different programs, emphasizing certain goals and subordinating others, even if all the alternative policies were in some general sense consistent with the congressional intent.

2. I have told this story in more detail in "The Birth of an Organization," chapter 16 in *Administrative Behavior,* 3rd ed. (New York: The Free Press, 1976).

As matters worked out, although vestigial organs representing each of the six approaches were still visible in the ECA after a year of operation, the balance of trade and European cooperation approaches generally prevailed, creating a measure of European economic stability and laying the groundwork for what later became the Common Market and ultimately the European Union. While each of the six approaches to the organization of ECA had some rational basis, trying to implement all of them simultaneously could (and almost did) create thorough confusion in the agency and among its clients. What was needed was not so much a "correct" conceptualization as one that could be understood by all the participants and that would facilitate action rather than paralyze it. The organization of ECA, as it evolved, provided a common problem representation within which all could work.

Finding the Limiting Resource

A second example illustrates the importance, in choosing a representation for a design problem, of identifying correctly the limiting resource or resources. A few years ago, the State Department was troubled by the congestion that affected its incoming communication lines whenever there was a crisis abroad. The teletypes, unable to output messages as rapidly as they were received, would fall many hours behind. Important messages to Washington were seriously delayed in transmission.

Since printing capacity was identified as the limiting factor, it was proposed to remedy the situation by substituting line printers for the teletypes, thereby increasing output by several orders of magnitude. No one asked about the next link in the chain: the capacity of officers at the country desks to process the messages that would come off the line printers. A deeper analysis would have shown that the real bottleneck in the process was the time and attention of the human decision makers who had to use the incoming information. Identification of the bottleneck would have generated in turn a more sophisticated design problem: How can incoming messages during a crisis be filtered in such a way that important information will have priority and will come to the attention of the decision makers, while unimportant information will be shunted aside until the crisis is past? Stated in this way, the design problem is not an easy one, but if a solution is found, even a partial one, it will at least tend to alleviate the real problem instead of aggravating it.

This is not an isolated example. The first generation of management information systems installed in large American companies were largely judged to have failed because their designers aimed at providing more information to managers, instead of protecting managers from irrelevant distractions of their attention.[3] A design representation suitable to a world in which the scarce factor is information may be exactly the wrong one for a world in which the scarce factor is attention.

As of the mid-1990s the lesson has still not been learned. An "information superhighway" is proclaimed without any concern about the traffic jams it can produce or the parking spaces it will require. Nothing in the new technology increases the number of hours in the day or the capacities of human beings to absorb information. The real design problem is not to provide more information to people but to allocate the time they have available for receiving information so that they will get only the information that is most important and relevant to the decisions they will make. The task is not to design information-distributing systems but intelligent information-filtering systems.[4]

Representations without Numbers

Many of the formal planning tools available to us call for representation of the design problem in quantitative form. Bayesian decision analysis, for example, requires that numerical utilities and "prior" probabilities be assigned to the possible decision outcomes and that "posterior" probabilities then be calculated for them on the basis of the estimated probability distributions of external events. With the assigned utilities and estimated probabilities in hand, the expected utility of each alternative can be computed and the best chosen.

Design problems often involve setting one or more parameters at values that will be neither too high nor too low. Such problems can often be conceptualized in economic terms as requiring the balancing of marginal

3. See H. A. Simon, *The New Science of Management Decision* (Englewood Cliffs, N.J.: Prentice-Hall, 1977), chapter 4.

4. H. A. Simon, "The Impact of Electronic Communications on Organizations, in R. Wolff (ed.), *Organizing Industrial Development* (Berlin: Walter de Gruyter, 1986).

benefits against marginal costs. Consider, for example, the task of regulating automobile emission standards.[5] The problem can be represented rationally as follows: (1) the quantity of emissions is a function of the number of cars, how far they are driven, and their design (hence cost); (2) the quality of air is a function of the level of emissions and of various geographical and meteorological parameters; (3) effects on human health depend on the quality of the air and the population exposed to it. An appropriate juxtaposition of these three functions produces a relation in which health is the dependent variable and the cost of automobiles the independent variable. If a dollar value is now assigned to health effects, all the ingredients will be present for a straightforward cost-benefit analysis of emission standards.

It is only necessary to state the problem in this way to show the preposterousness of attempting such calculations. Nevertheless, when this problem was presented to the National Academy of Sciences—not because it was solvable, but because the Congress had to make a decision about emission standards—the conceptual scheme for cost-benefit analysis proved to be an excellent representation for organizing the subcommittees of experts who were asked to contribute their advice. One subcommittee, mainly of engineers, examined the cost of redesigning cars to reduce emissions. A second committee, experts in atmospheric chemistry and meteorology, analysed the relations between emissions and air quality. A whole set of committees of medical experts reviewed the evidence on the health effects of the principal pollutants. Yet another committee, staffed by economists, undertook to make estimates of the values that should be assigned to health effects.

None of these committees was able to arrive at estimates that were believable in more than an order-of-magnitude sense, unless they were the estimates of auto costs which might have been accurate within a factor of two. In general the medical committees were unwilling or unable to make any marginal estimates at all, confining themselves to finding threshold levels of air quality at which the health effects of pollutants were

5. Coordinating Committee on Air Quality Studies, National Academy of Sciences and National Academy of Engineering, *Air Quality and Automobile Emission Control*, Vols. 1–4, no. 93–23 (Washington: Government Printing Office, 1974).

detectable. Given these kinds of findings and assessments, there was no way in which the hypothetical cost-benefit analysis scheme could be applied literally. Nevertheless the scheme provided a conceptual framework in which the findings could be related to each other, and in which the coordinating committee that had to assemble the pieces of the puzzle could judge the reasonableness of proposed standards. Even in this complex setting reasonable men could set upper and lower bounds on emission levels that, if not dictated by the evidence, at least were consistent with it. If optimizing was out of the question, the framework allowed the committee to arrive at a satisficing decision that was not outrageous or indefensible.

One may regard "defensibility" as a weak standard for a decision on a matter as consequential as automobile emissions. But it is probably the strictest standard we can generally satisfy with real-world problems of this complexity. Even in situations of this kind (perhaps it would be better to say "especially in situations of this kind") an appropriate representation of the problem may be essential to organizing efforts toward solution and to achieving some kind of clarity about how proposed solutions are to be judged. Numbers are not the name of this game but rather representational structures that permit functional reasoning, however qualitative it may be.

Data for Planning

If, given a good problem representation, rational analysis can sometimes be carried out even in the absence of most of the relevant numbers, still we should not make a virtue of this necessity. The quality of design is likely to depend heavily on the quality of the data available. The task is not to design without data but to incorporate assessments of the quality of the data, or its lack of quality, in the design process itself. Setting automobile emission standards may call for a different approach to data than calculating the optimal profile for an airplane wing.

What paths are open to us when we must plan in the face of extremely poor data? One minimal strategy, which scientists have generally followed for several hundred years but planners sometimes ignore, is to associate with every estimated quantity a measure of its precision. Labeling estimates in this way does not make them more reliable, but it does remind us how hard or soft they are and hence how much trust to place in them.

Prediction

Data about the future—predictions—are commonly the weakest points in our armor of fact. Good predictions have two requisites that are often hard to come by. First, they require either a theoretical understanding of the phenomena to be predicted, as a basis for the prediction model, or phenomena that are sufficiently regular that they can simply be extrapolated. Since the latter condition is seldom satisfied by data about human affairs (or even about the weather), our predictions will generally be only as good as our theories.

The second requisite for prediction is having reliable data about the initial conditions—the starting point from which the extrapolation is to be made. Systems vary in the extent to which their paths are sensitive to small changes in initial conditions. Weather prediction is difficult in good part because the course of meteorological events is highly sensitive to the details of initial conditions. We have every reason to think that social phenomena are similarly sensitive.

Since the consequences of design lie in the future, it would seem that forecasting is an unavoidable part of every design process. If that is true, it is cause for pessimism about design, for the record in forecasting even such "simple" variables as population is dismal. If there is any way to design without forecasts, we should seize on it.

Consider the much-discussed Club of Rome report, which predicted a twenty-first century doomsday of overpopulation, resource exhaustion, and famine.[6] Since the specifics of the model used to make the Club of Rome predictions have already been much criticized, I don't have to examine those specifics here. My point is a broader one. The Club of Rome report predicted both too much and too little. It predicted too *much*, because its specific doomsday dates are not believable, and if believable, would not be important. We do not want to know when disaster is going

6. Donella Meadows et al., *The Limits to Growth* (N.Y.: Universe Books, 1972).

to strike but how to avoid it. Without any specific predictions we know that a system with exponential population growth and limited resources will sooner or later come to some bad end. For planning purposes we wish only to have some sense of the time scale of events, to know at least whether we are talking about years, decades, generations, or centuries. For most design purposes that is as much prediction as we need.

The Club of Rome report predicted too *little* because it emphasized a single possible time path rather than focusing upon alternative futures. The heart of the data problem for design is not forecasting but constructing alternative scenarios for the future and analyzing their sensitivity to errors in the theory and data. What is said here about environmental modeling applies as well to efforts aimed specifically at modeling climate change caused by global warming. Predicting the exact course of global warming is a thankless task. Much more feasible and useful is generating alternative policies which can be introduced at the appropriate times for slowing the warming, mitigating its unfavorable effects and taking advantage of favorable effects.[7]

How can we go about designing an acceptable future for the energy and environmental needs of a society? First, we select some planning horizons: perhaps five years for short-term plans, a generation for middle-term plans, and a century or two for long-term plans. There is no need to construct detailed forecasts for each of these time perspectives. Instead we can concentrate our analytic resources on examining alternative target states for the system for the short, middle, and long run. By a target state I mean upper bounds on the quantities of energy used and pollutants produced. Having chosen a desirable (or acceptable) target state, and having satisfied ourselves that its realizability is not unduly sensitive to unpredictables, we can then turn our attention to constructing paths that lead from the present to that desired future.

Design for distant futures would be wholly impossible if remote events had to be envisioned in detail. What makes such design even conceivable is that we need to know or guess about the future only enough to guide the commitments we must make today. Future contingencies that have no

7. H. A. Simon, "Prediction and Prescription in Systems Modeling," *Operations Research*, 38(1990):7–14.

implications for present commitment have no relevance to design. I will have more to say on this point presently.

Feedback

Few of the adaptive systems that have been forged by evolution or shaped by man depend on prediction as their main means for coping with the future. Two complementary mechanisms for dealing with changes in the external environment are often far more effective than prediction: homeostatic mechanisms that make the system relatively insensitive to the environment and retrospective feedback adjustment to the environment's variation.

Thus a stock of inventories permits a factory to operate without concern for very short-run fluctuations in product orders. Energy storage in the tissues of a predator enables it to cope with uncertainties in the availability of prey. A modest excess of capacity in electric generating plants avoids the need for precise estimation of peak loads. Homeostatic mechanisms are especially useful for handling short-range fluctuations in the environment, hence for making short-range prediction unnecessary.

Feedback mechanisms, on the other hand, by continually responding to discrepancies between a system's actual and desired states, adapt it to long-range fluctuations in the environment without forecasting. In whatever directions the environment changes, the feedback adjustment tracks it, with of course some delay.

In domains where some reasonable degree of prediction is possible, a system's adaptation to its environment can usually be improved by combining predictive control with homeostatic and feedback methods. It is well known in control theory, however, that active, feedforward control, using predictions, can throw a system into undamped oscillation unless the control responses are carefully designed to maintain stability. Because of the possible destabilizing effects of taking inaccurate predictive data too seriously, it is sometimes advantageous to omit prediction entirely, relying wholly on feedback, unless the quality of the predictions is high.[8]

8. The design of dynamic programming schemes that use a combination of prediction and feedback to control factory systems is discussed in Holt, Modigliani, Muth, and Simon, *Planning, Production, Inventories, and Work Force* (Englewood Cliffs, N.J., Prentice-Hall, 1960).

Who is the Client?

It may seem peculiar to ask, "Who is the client?" when speaking of the design of large social systems. The question need not be raised about smaller-scale design tasks, since the answer is built into the definitions of the professional roles of designers. At microsocial levels of design it is tacitly assumed that the professional architect, attorney, civil engineer, or physician works for a specified client and that the needs and wishes of the client determine the goals of the professional's work. In this model of professional activity the architect designs a house that meets the living requirements of the client, while the physician plans a course of treatment for the patient's ailments. Although in practice matters are not so simple, this definition of the professional role greatly facilitates the development of technologies for each of the professions, for it means that consequences going beyond the client's goals don't have to enter into the design calculations. The architect need not decide if the funds the client wants to spend for a house would be better spent, from society's standpoint, on housing for low-income families. The physician need not ask whether society would be better off if the patient were dead.

Thus the traditional definition of the professional's role is highly compatible with bounded rationality, which is most comfortable with problems having clear-cut and limited goals. But as knowledge grows, the role of the professional comes under questioning. Developments in technology give professionals the power to produce larger and broader effects at the same time that they become more clearly aware of the remote consequences of their prescriptions.

In part this complication in the professional's role comes about simply as a direct by-product of the growth of knowledge. Whether through the modification of professional norms or through direct intervention of government, new obligations are placed on the professional to take account of the external effects—the consequences beyond the client's concern—that are produced by the designs.

These same developments cause the professional to redefine the concept of the client. The psychiatrist working with an individual patient becomes a family counselor. The engineer begins to take into account the environmental impact of new products. Finally, as the society and its central government take on a wider range of responsibilities, more and more professionals find they no longer serve individual clients but are employed directly by agencies of the state. Almost all of the professions today are undergoing self-examination as they experience the pressures generated by these complications in their roles. Architecture, medicine, and engineering all exhibit the stresses engendered by this process.

Professional-Client Relations

Architects are especially conflicted for several reasons. First, they have always assigned themselves the dual role of artist and professional, two roles that often make inconsistent demands. As artists they wish to realize esthetic goals that may be quite independent of clients' expressed or understood desires. If a client comports himself as an (idealized) Renaissance patron, there may be no difficulty, for the patron does not impose his views of beauty on the artist. But if the client's approach to building takes a more utilitarian bent, and he is not willing to sacrifice what he conceives as usefulness for what the architect conceives as beauty, then the relation between them may be tainted with mistrust and deception. At best, the architect becomes teacher and advocate, not simple executor of the client's purposes.

I once asked Mies van der Rohe, then my faculty colleague at Illinois Institute of Technology, how he got the opportunity to build the Tugendhat house—a startlingly modern design at the time of its construction. The prospective owner had come to Mies after seeing some of the quite conventional houses he had earlier designed in the Netherlands when he was still an apprentice. "Wasn't the client shocked," I asked, "when you put before him your glass and metal design?" "Yes," said Mies, viewing the tip of his cigar reflectively, "he wasn't very happy at first. But then we smoked some good cigars, . . . and we drank some glasses of a good Rhein wine, . . . and then he began to like it very much."

A second and increasingly acute problem for architects is that, when they take on the task of designing whole complexes or areas instead of single buildings, their professional training does not provide them with clear design criteria. In city planning, for example, the boundary between the design of physical structures and the design of social systems dissolves almost completely. Since there is little in the knowledge base or portfolio of techniques of architecture that qualifies the professional to plan such social systems, the approach to the design tends to be highly idiosyncratic, reflecting little that can be described as professional consensus, and even less that can be described as empirically based analytic technique.

In the medical profession the stresses take slightly different forms. The first arises from the resource allocation problem—balancing the cost of medical care against its quality. Traditionally patients got such care as they could afford, or as the doctor could afford to provide for them—one can look at it either way. Today with indirect channels of payment for most medical services, budget constraints are harder to define and monitor, and ethical choices have to be made explicitly that formerly were made tacitly.

The second stress in the design of medical care and treatments derives from the advance of medical technology, which gives the physician a degree of control over life and death that is vastly greater than in the past. And so the traditional view, which opts unconditionally for life, no longer remains unquestioned. Even harder questions arise with new technical means for modifying genetic processes and for manipulating the mind. In the traditional professional-client relation, the client's needs and wants are given. The environment (including the functioning of the body) is to be adapted to the client's goals, not the goals to the environment. Yet much utopian thought has conceived of change in both directions. Society was to be made more fit for human habitation, but the human inhabitants were also to be modified to make them more fit for society. Today we are deeply conflicted about how far we should go in "improving" human beings involuntarily. The movie *The Clockwork Orange* states the conflict dramatically by asking whether we are justified in destroying the capability for willful action even to prevent viciousness.

The case of the engineer presents yet another aspect of the problems that growing technical power and growing awareness of remote consequences bring. Most engineering is done within the context of business and governmental organizations. In this environment there is continuing potential for conflict between the decision criteria defined by the profession and those enforced by the organization. In the hypothetical business firms of the pure theory of competition, discussed in chapter 2, organizational criteria would prevail. In the more complex world in which we actually live, the professional engineers possess substantial discretion to give professional considerations priority over the goals of the organization. If they choose to exercise that discretion, they must decide who the client is. In particular they must decide which of the positive and negative externalities to which the artifacts they are designing will give rise should be incorporated in the design criteria.

Society as the Client

It may seem obvious that all ambiguities should be resolved by identifying the client with the whole society. That would be a clear-cut solution in a world without conflict of interest or uncertainty in professional judgment. But when conflict and uncertainty are present, it is a solution that abdicates organized social control over professionals and leaves it to them to define social goals and priorities. If some measure of control is to be maintained, the institutions of the society must share with the professional the redefinition of the goals of design.

The client seeks to control professionals not only by defining their goals of design but also by reacting to the plans they propose. It is well known that physicians' patients fail to take much of the medicine that is prescribed for them. Society as client is no more docile than are medical patients. In any planning whose implementation involves a pattern of human behavior, that behavior must be motivated. Knowledge that "it is for your own good" seldom provides adequate motivation.

The members of an organization or a society for whom plans are made are not passive instruments, but are themselves designers who are seeking to use the system to further their own goals. Organization theory deals with this motivational question by examining organizations in terms of the balance between the inducements that are provided to members to perform their organizational roles and the contributions that the members thereby provide to the achievement of organizational goals.[9]

A not dissimilar representation of the social planning process views it as a game between the planners and those whose behavior they seek to influence. The planners make their move (i.e., implement their design),

9. The notion of organizational survival and equilibrium depending on the balance of inducements and contributions is due to Chester I. Barnard, *The Functions of the Executive* (Cambridge: Harvard University Press, 1938).

and those who are affected by it then alter their own behavior to achieve their goals in the changed environment. The gaming aspects of social planning are particularly evident in the domain of economic stabilization policies, where the adaptive response of firms and consumers to monetary and fiscal policies may largely neutralize or negate those policies. The claims of monetarists, and especially of the "rational expectations" theorists, that government is helpless to influence employment levels by using the standard Keynesian tools of monetary and fiscal policy and that attempts to reduce unemployment can only cause inflation, are based on the assumption that public responses to these measures will be strongly and rapidly adaptive.

Except for economics it is still relatively rare for social planning and policy discussions to include in any systematic way the possible "gaming" responses to plans. For example, until quite recently it was common to design new urban transit facilities without envisioning the possible relocations of population within the urban area that would be produced by the new facilities themselves. Yet such effects have been known and observed for half a century. Social planning techniques need to be expanded to encompass them routinely.

Organizations in Social Design

In introducing the subject of social design, I used the Constitution of the United States as an example. Configuring organizations, whether business corporations, governmental organizations, voluntary societies or others, is one of society's most important design tasks. If we human beings were isolated monads—small, hermetically sealed particles that had no mutual relations except occasional elastic collisions—we would not have to concern ourselves with the design of organizations. But, contrary to libertarian rhetoric, we are not monads. From birth until death, our ability to reach our goals, even to survive, is tightly linked to our social interactions with others in our society.

The rules imposed upon us by organizations—the organizations that employ us and the organizations that govern us—restrict our liberties in a variety of ways. But these same organizations provide us with opportunities for reaching goals and attaining freedoms that we could not even imagine reaching by individual effort. For example, almost everyone who will read these lines has an income that is astronomical by comparison with the world average. If we were to assign a single cause to our good fortune, we would have to attribute it to being born in the right place at the right time: in a society that is able to maintain order (through public organizations), to produce efficiently (largely through business organizations), and to maintain the infrastructure required for high production (again largely through public organizations). We have even discovered, in our society and a modest number of others, how to design organizations, business and governmental, that do not interfere egregiously with our freedoms, including those of speech and thought.

This is not the place to enter into a long disquisition on organizational design, private and public, which has a large literature of its own.[10] But one can hardly pass by governments and business firms in complete silence in a chapter on the design of social structures. A society's organizations are matters not only of specialized professional concern but of broad public concern.

Today, organizations, and especially governmental organizations, have an exceedingly bad press in our society. The terms "politician" and "bureaucrat" are not used as descriptors but as pejoratives. While the events in Oklahoma City surely did not evoke public approval, the general horrified reaction was not to the anti-governmental attitudes that the bombing expressed but to the killings. There is more than a little anarchism (usually phrased as libertarianism) in the current American credo (and for that matter, in our credo since the time of the Founding Fathers).

Organizational design, then, is a matter for urgent attention in any curriculum on social design. Organizations are exceedingly complex systems that share many properties with other complex systems—for example, their typically hierarchical structure. Questions of organizational design

---

10. My views on some of these matters have been expounded at length in H. A. Simon, *Administrative Behavior, 3rd ed.,* (New York, NY: The Free Press, 1976); H. A. Simon, V. A. Thompson and D. W. Smithburg, *Public Administration* (New Brunswick, NJ: Transaction Publishers, 1991); and J. G. March and H. A. Simon, *Organizations, 2nd ed.,* (Cambridge, MA: Blackwell, 1993). On the nature of business organizations, and especially the role of organizational identification in maintaining them, see chapter 2 of the present volume.

will reappear, from time to time, as part of the discussion of complex systems in chapters 7 and 8, below, and especially in connection with the use of hierarchy and "near decomposability" as a basis for specialization.

Time and Space Horizons for Design

Each of us sits in a long dark hall within a circle of light cast by a small lamp. The lamplight penetrates a few feet up and down the hall, then rapidly attenuates, diluted by the vast darkness of future and past that surrounds it.

We are curious about that darkness. We consult soothsayers and forecasters of the economy and the weather, but we also search backward for our "roots." Some years ago I conducted such a search in the Rheinland villages near Mainz where my paternal ancestors had lived. I found records of grandparents readily enough, and even of great-grandparents and beyond. But before I had gone far—scarcely back to the 18th century—I came to the edge of the circle of light. Darkness closed in again in the little towns of Ebersheim, Woerstadt, and Partenheim, and I could see no farther back.

History, archaeology, geology, and astronomy provide us with narrow beams that penetrate immense distances down the hallway of the past but illuminate it only fitfully—a statesman or philosopher here, a battle there, some hominoid bones buried with pieces of chipped stone, fossils embedded in ancient rock, rumors of a great explosion. We read about the past with immense interest. A few spots caught by the beams take on a vividness and immediacy that capture, for a moment, our attention and our hearts—some Greek warriors camped before Troy, a man on a cross, the painted figure of a deer glimpsed by flickering torchlight on the wall of a limestone cave. But mostly the figures are shadowy, and our attention shifts back to the present.

The light dims even more rapidly in the opposite direction, toward the future. Although we are titillated by Sunday Supplement descriptions of a cooling Sun, it is our own mortality, just a few years away, and not the Earth's, with which we are preoccupied. We can empathize with parents and grandparents whom we have known, or of whom we have had first-hand accounts, and in the opposite direction with children and grandchildren. But beyond that circle our concern is more curious and intellectual than emotional. We even find it difficult to define which distant events are the triumphs and which the catastrophes, who the heroes and who the villains.

Discounting the Future

Thus the events and prospective events that enter into our value systems are all dated, and the importance we attach to them generally drops off sharply with their distance in time. For the creatures of bounded rationality that we are, this is fortunate. If our decisions depended equally upon their remote and their proximate consequences, we could never act but would be forever lost in thought. By applying a heavy discount factor to events, attenuating them with their remoteness in time and space, we reduce our problems of choice to a size commensurate with our limited computing capabilities. We guarantee that, when we integrate outcomes over the future and the world, the integral will converge.

Sociobiologists, in their analyses of egoism and altruism, undertake to explain how the forces of evolution would necessarily produce organisms more protective of their offspring and their kin than of unrelated creatures. This evolutionary account does not explain, however, why the concern tends to be so myopic with respect to the future. At least one part of the explanation is that we are unable to think coherently about the remote future, and particularly about the distant consequences of our actions. Our myopia is not adaptive, but symptomatic of the limits of our adaptability. It is one of the constraints on adaptation belonging to the inner environment.

The economist expresses this discounting of the future by a rate of interest. To find the present value of a future dollar, he applies, backwards, a compound discount rate that shrinks the dollar by a fixed percentage for each step from the present. Even a moderate rate of interest can make the dollars of the next century look quite inconsequential for our present decisions. There is a vast literature seeking to explain, none too convincingly, what determines the time rate of discount used by savers. (In modern times it has hovered remarkably steadily around 3 percent per annum, after appropriate adjustment for risk and inflation.) There is also a considerable literature seeking to determine what the social rate of interest *should* be—what the rate of exchange should be between the welfare of this generation and the welfare of its descendants.

The rate of interest should not be confused with another factor that discounts the importance of the future with respect to the present. Even if we are aware of certain unfavorable events that will occur in the distant future, there may be nothing to be done about them today. If we knew that the wheat harvest was going to fail in the year 2020, we would be ill-advised to store wheat now. Our unconcern with a distant future is not merely a failure of empathy but a recognition that (1) we shall probably not be able to foresee and calculate the consequences of our actions for more than short distances into the future and (2) those consequences will in any case be diffuse rather than specific.

The important decisions we make about the future are primarily the decisions about spending and saving—about how we shall allocate our production between present and future satisfactions. And in saving, we count flexibility among the important attributes of the objects of our investment, because flexibility insures the value of those investments against the events that will surely occur but which we cannot predict. It will (or should) bias our investments in the direction of structures that can be shifted from one use to another, and to knowledge that is fundamental enough not soon to be outmoded—knowledge that may itself provide a basis for continuing adaptation to the changing environment.

The Change in Time Perspective

One of the noteworthy characteristics of our century is the shift that appears to be taking place, especially in the industrialized world, in our time perspectives. For example, embedded in the energy-environment problem that confronts us today, we can see three almost independent aspects. The first is our immediate dependence on petroleum, which we must reduce to protect ourselves from political blackmail and to achieve a balance of international payments. The second is the prospect of the exhaustion of oil and gas supplies, a problem that must be solved within about a generation, mostly by the use of coal and nuclear energy. The third is the joint problem of the exhaustion of fossil fuels and the impact of their combustion on the climate. The time scale of this third problem is a century or so.

What is remarkable in our age, and relatively novel I believe, is the amount of attention we pay to the third problem. Perhaps it is just that we have all three confused in our minds and have not sorted them out to the point where we can think about the more pressing ones without concern for the other. But I do not think that is the reason. I believe there has been a genuine downward shift in the social interest rate we apply to discount events that are remote in time and space.

There are some obvious reasons for our new concern with matters that are remote in time and space. Among these are the relatively new facts of instantaneous worldwide communication and rapid air transportation. Consequent on these is the continually increasing economic and military interdependence of all the nations. More subtle than either of these causes is the progress of human knowledge, especially of science. I have already commented on the way in which archaeology, geology, anthropology, and cosmology have lengthened our perspectives. But in addition new laboratory technologies have vastly increased our ability to detect and assess small and indirect effects of our actions. Oscar Wilde once claimed that there were no fogs on the River Thames until Turner, by painting them, revealed them to the residents of London. In the same way our atmosphere contained no noxious substances in quantities of a few parts per million until chromatography and other sensitive analytic techniques showed their presence and measured them. DDT was an entirely beneficent insecticide until we detected its presence in falcons' eggs and in fish. If eating the apple revealed to us the nature of good and evil, modern analytic tools have taught us how to detect good and evil in minute amounts and at immense distances in time and space.

It may be objected that there has been no such lengthening of social time perspectives as I have claimed. What perspective can be longer than the eternity of life after death that is so central to Christian thought, or longer than the repeated reincarnations of Eastern religions? But the attitudes toward the future engendered by those beliefs are very different from the ones I have been discussing. The future with which the Christian is concerned is his own future in the light of his current conduct. There is nothing in the belief in an afterlife or a reincarnation that calls attention to the future consequences for this world of one's present actions. Nor do I find in those religious beliefs anything resembling the contemporary concern for the fragility of the environment on which human life depends or for the power of human actions to make that environment more or less habitable in the future. It does appear therefore that there has been a genuine shift in our orientation to time and a significant lengthening in time perspectives.

Defining Progress

As the web of cause and effect is woven tighter, we put severe loads upon our planning and decision-making procedures to deal with these remote effects. There is a continuing race between the part of our new science and knowledge that enables us to see more distant views and the part that enables us to deal with what we see. And if we live in a time that is sometimes pessimistic about technology, it is because we have learned to look farther than our arms can reach.

Defining what is meant by progress in human societies is not easy. Increasing success in meeting basic human needs for food, shelter, and health is one kind of definition that most people would agree upon. Another would be an average increase in human happiness. With the advance of productive technology, we can claim that there has been major progress by the first criterion; but what has been said in chapter 2 about changing aspiration levels would lead us to doubt whether progress is possible if we use the second criterion, human happiness, to measure it. There is no reason to suppose that a modern industrial society is more conducive to human happiness than the simpler, if more austere, societies that preceded it. On the other hand, there seems to be little empirical basis for the nostalgia that is sometimes expressed for an imagined (and imaginary) happier or more humane past.

A third way of measuring progress is in terms of intentions rather than outcomes—what might be called moral progress. Moral progress has always been associated with the capacity to respond to universal values—to grant equal weight to the needs and claims of all mankind, present and future. It can be argued that the growth of knowledge of the kinds I have been describing represents such moral progress.

But we should not be hasty in our evaluation of the consequences of lengthening perspectives in space or time. The present century is not lacking in horrible examples of man's inhumanity to man. We must be alert also to the possibility that rationality applied to a broader domain will simply be a more calculatedly rational selfishness than the impulsive selfishness of the past.

The Management of Attention

From a pragmatic standpoint we are concerned with the future because securing a satisfactory future may require actions in the present. Any interest in the future that goes beyond this call for present action has to be charged to pure curiosity. It belongs to our recreational rather than our working day. Thus our present concern for the short-run energy problem is quite different from our concern for the long-run problem or even the middle-run problem. The actions we have to take today, if we are to improve the short-run situation, are largely actions that will reduce our use of energy—there are only modest prospects of a substantial short-run increase in supply. The actions we have to take with respect to the middle-run problem are largely actions on a large scale toward the development and exploitation of some mix of technologies for the conversion of coal, mining of oil sands and shales, and safe nuclear fission or fusion. The principal actions we can take now with respect to the long-range energy problem are primarily knowledge-acquiring actions—research programs to develop nuclear fusion and solar technologies and to gain a deeper understanding of the environmental consequences of all the alternatives.

The energy problem is rather typical in this respect of large-scale design problems. In addition to the things we can do to produce immediate consequences, we must anticipate the time lags involved in developing new capital plant and the even greater time lags involved in developing the body of technology and other knowledge that we will need in the more distant future. Attention of the decision-making bodies has to be allocated correspondingly.

It is a commonplace organizational phenomenon that attending to the needs of the moment—putting out fires—takes precedence over attending to the needs for new capital investment or new knowledge. The more crowded the total agenda and the more frequently emergencies arise, the more likely it is that the middle-range and long range decisions will be neglected. In formal organizations a remedy is often sought for this condition by creating planning groups that are insulated in one way or another from the momentary pressures upon the organization. Planning units face two hazards. On the one hand, and especially if they are competently staffed, they may be consulted more and more frequently for help on immediate problems until they are sucked into the operating organization and can no longer perform their planning functions. If they are sufficiently well sealed off from the rest of the organization to prevent this from happening, then they may find the reverse channel blocked—they may be unable to influence decisions in the operating organization. There is no simple or automatic way to remove these difficulties once and for all. They require repeated attention from the organization's leadership.

Designing without Final Goals

To speak of planning without goals may strike one as a contradiction in terms.[11] It seems "obvious" that the very concept of rationality implies goals at which thought and action are aimed. How can we evaluate a design unless we have well-defined criteria against which to judge it, and how can the design process itself proceed without such criteria to guide it?

Some answer has already been given to these questions in chapter 4, in the discussion of discovery processes. We saw there that search guided by only the most general heuristics of "interestingness" or novelty is a fully realizable activity. This kind of search, which provides the mechanism for scientific discovery, may also provide the most suitable model of the social design process.

It is generally recognized that in order to acquire new tastes in music, a good prescription is to hear more music; in painting, to look at paintings; in wine, to drink good wines. Exposure to new experiences is almost certain to change the criteria of choice, and most human beings deliberately seek out such experiences.

A paradoxical, but perhaps realistic, view of design goals is that their function is to motivate activity which in turn will generate new goals. For example, when about fifty years ago an extensive renewal program was

11. This section owes much to James G. March, who has thought deeply on these lines. See his "Bounded Rationality, Ambiguity, and the Engineering of Choice," *Bell Journal of Economics* 9(1978):587–608.

begun in the city of Pittsburgh, a principal goal of the program was to rebuild the center of the city, the so-called Golden Triangle. Architects have had much to say, favorable and unfavorable, about the esthetic qualities of the plans that were carried out. But such evaluations are largely beside the point. The main consequence of the initial step of redevelopment was to demonstrate the possibility of creating an attractive and functional central city on this site, a demonstration that was followed by many subsequent construction activities that have changed the whole face of the city and the attitudes of its inhabitants.

It is also beside the point to ask whether the later stages of the development were consistent with the initial one—whether the original designs were realized. Each step of implementation created a new situation; and the new situation provided a starting point for fresh design activity.

Making complex designs that are implemented over a long period of time and continually modified in the course of implementation has much in common with painting in oil. In oil painting every new spot of pigment laid on the canvas creates some kind of pattern that provides a continuing source of new ideas to the painter. The painting process is a process of cyclical interaction between painter and canvas in which current goals lead to new applications of paint, while the gradually changing pattern suggests new goals.

The Starting Point

The idea of final goals is inconsistent with our limited ability to foretell or determine the future. The real result of our actions is to establish initial conditions for the next succeeding stage of action. What we call "final" goals are in fact criteria for choosing the initial conditions that we will leave to our successors.

How do we want to leave the world for the next generation? What are good initial conditions for them? One desideratum would be a world offering as many alternatives as possible to future decision makers, avoiding irreversible commitments that they cannot undo. It is the aura of irreversibility hanging about so many of the decisions of nuclear energy deployment that makes these decisions so difficult.

A second desideratum is to leave the next generation of decision makers with a better body of knowledge and a greater capacity for experience.

The aim here is to enable them not just to evaluate alternatives better but especially to experience the world in more and richer ways.

Becker and Stigler have argued that considerations of the sort I have been advancing can be accommodated without giving up the idea of fixed goals.[12] All that is required, they say, is that the utilities to be obtained from actions be defined in sufficiently abstract form. In their scheme the utility yielded by an hour's listening to music increases with one's capacity for musical enjoyment, and this capacity is a kind of capital that can be increased by a prior investment in musical listening. While I find their way of putting the matter a trifle humorless, perhaps it makes the idea of rational behavior without goals less mysterious. If we conceive human beings as having some kind of alterable capacity for enjoyment and appreciation of life, then surely it is a reasonable goal for social decision to invest in that capacity for future enjoyment.

Designing as Valued Activity

Closely related to the notion that new goals may emerge from creating designs is the idea that one goal of planning may be the design activity itself. The act of envisioning possibilities and elaborating them is itself a pleasurable and valuable experience. Just as realized plans may be a source of new experiences, so new prospects are opened up at each step in the process of design. Designing is a kind of mental window shopping. Purchases do not have to be made to get pleasure from it.

One of the charges sometimes laid against modern science and technology is that if we know *how* to do something, we cannot resist doing it. While one can think of counterexamples, the claim has some measure of truth. One can envisage a future, however, in which our main interest in both science and design will lie in what they teach us about the world and not in what they allow us to do to the world. Design like science is a tool for understanding as well as for acting.

12. G. J. Stigler and G. S. Becker, "De Gustibus non est Disputandum," American Economic Review, 67(1977):76–90.

Social Planning and Evolution

Social planning without fixed goals has much in common with the processes of biological evolution. Social planning, no less than evolution, is myopic. Looking a short distance ahead, it tries to generate a future that is a little better (read "fitter") than the present. In so doing, it creates a new situation in which the process is then repeated. In the theory of evolution there are no theorems that extract a long-run direction of development from this myopic hill climbing. In fact evolutionary biologists are extremely wary of postulating such a direction or of introducing any notion of "progress." By definition the fit are those who survive and multiply.

Whether there is a long-run direction in evolution, and whether that direction is to be considered progress are of course two different questions. We might answer the former affirmatively but the latter negatively. Let me venture a speculation about the direction of social and biological evolution, which I will develop further in the next two chapters. My speculation is emphatically *not* a claim about progress.

From a reading of evolutionary history—whether biological or social—one might conjecture that there has been a long-run trend toward variety and complexity. There are more than a hundred kinds of atoms, thousands of kinds of inorganic molecules, hundreds of thousands of organic molecules, and millions of species of living organisms. Mankind has elaborated several thousand distinct languages, and modern industrial societies count their specialized occupations in the tens of thousands.

I shall emphasize in the following chapters that forms can proliferate in this way because the more complex arise out of a combinatoric play upon the simpler. The larger and richer the collection of building blocks that is available for construction, the more elaborate are the structures that can be generated.

If there is such a trend toward variety, then evolution is not to be understood as a series of tournaments for the occupation of a fixed set of environmental niches, each tournament won by the organism that is fittest for that niche. Instead evolution brings about a proliferation of niches. The environments to which most biological organisms adapt are formed mainly of other organisms, and the environments to which human beings adapt, mainly of other human beings. Each new bird or mammal provides a niche for one or more new kind of flea.

Vannevar Bush wrote of science as an "endless frontier." It can be endless, as can be the process of design and the evolution of human society, because there is no limit on diversity in the world. By combinatorics on a few primitive elements, unbounded variety can be created.

The Curriculum for Social Design

Our examination of the social planning process here suggests some extension of the curriculum for design that was proposed in the last chapter. Topic 7, the representation of design problems, must be expanded to incorporate the skills of constructing organizations as frameworks for problem representation, building representations around limiting factors, and representing non-numerical problems. Our discussion also suggests at least six new topics for the curriculum:

1. *Bounded rationality.* The meaning of rationality in situations where the complexity of the environment is immensely greater than the computational powers of the adaptive system.

2. *Data for planning.* Methods of forecasting, the use of prediction and feedback in control.

3. *Identifying the client.* Professional-client relations, society as the client, the client as player in a game.

4. *Organizations in social design.* Not only is social design carried out mainly by people working in organizations, but an important goal of the design is to fashion and change social organization in general and individual organizations in particular.

5. *Time and space horizons.* The discounting of time, defining progress, managing attention.

6. *Designing without final goals.* Designing for future flexibility, design activity as goal, designing an evolving system.

With the exception of control theory and game theory, which are of central importance to topics 2 and 3, the design tools relevant to these additional topics are in general less formal than those we described in the previous chapter. But whether we have the formal tools we need or not, the topics are too crucial to the social design process to permit them to be ignored or omitted from the curriculum.

Our age is one in which people are not reluctant to express their pessimism and anxieties. It is true that humanity is faced with many problems. It always has been but perhaps not always with such keen awareness of them as we have today. We might be more optimistic if we recognized that we do not have to solve all of these problems. Our essential task—a big enough one to be sure—is simply to keep open the options for the future or perhaps even to broaden them a bit by creating new variety and new niches. Our grandchildren cannot ask more of us than that we offer to them the same chance for adventure, for the pursuit of new and interesting designs, that we have had.

7

Alternative Views of Complexity

The preceding chapters of this book have discussed several kinds of artificial systems. The examples we have examined—in particular, economic systems, the business firm, the human mind, sophisticated engineering designs, and social plans—range from the moderately to the exceedingly complex (not necessarily in the order in which I have just listed them). These final two chapters address the topic of complexity more generally, to see what light it casts on the structure and operation of these and other large systems that are prominent in our world today.

Conceptions of Complexity

This century has seen recurrent bursts of interest in complexity and complex systems. An early eruption, after World War I, gave birth to the term "holism," and to interest in "Gestalts" and "creative evolution." In a second major eruption, after World War II, the favorite terms were "information," "feedback," "cybernetics," and "general systems." In the current eruption, complexity is often associated with "chaos," "adaptive systems," "genetic algorithms," and "cellular automata."

While sharing a concern with complexity, the three eruptions selected different aspects of the complex for special attention. The post-WWI interest in complexity, focusing on the claim that the whole transcends the sum of the parts, was strongly anti-reductionist in flavor. The post-WWII outburst was rather neutral on the issue of reductionism, focusing on the roles of feedback and homeostasis (self-stabilization) in maintaining complex systems. The current interest in complexity focuses mainly on mechanisms that create and sustain complexity and on analytic tools for describing and analyzing it.

Holism and Reductionism

"Holism" is a modern name for a very old idea. In the words of its author, the South African statesman and philosopher, J. C. Smuts:

[Holism] regards natural objects as wholes. . . . It looks upon nature as consisting of discrete, concrete bodies and things . . . [which] are not entirely resolvable into parts; and . . . which are more then the sums of their parts, and the mechanical putting together of their parts will not produce them or account for their characters and behavior.[1]

Holism can be given weaker or stronger interpretations. Applied to living systems, the strong claim that "the putting together of their parts will not produce them or account for their characters and behaviors" implies a vitalism that is wholly antithetical to modern molecular biology. Applied to minds in particular, it is used to support both the claim that machines cannot think and the claim that thinking involves more than the arrangement and behavior of neurons. Applied to complex systems in general, it postulates new system properties and relations among subsystems that had no place in the system components; hence it calls for emergence, a "creative" principle. Mechanistic explanations of emergence are rejected.

In a weaker interpretation, emergence simply means that the parts of a complex system have mutual relations that do not exist for the parts in isolation. Thus, there can be gravitational attractions among bodies only when two or more bodies interact with each other. We can learn something about the (relative) gravitational accelerations of binary stars, but not of isolated stars.

In the same way, if we study the structures only of individual proteins, nothing presages the way in which one protein molecule, serving as an enzyme, can provide a template into which two other molecules can insert themselves and be held while they undergo a reaction linking them. The template, a real enough physical property of the enzyme, has no function until it is placed in an environment of other molecules of a certain kind.

1. J. C. Smuts, "Holism," *Encyclopaedia Britannica,* 14th ed., vol. 11 (1929), p. 640.

Even though the template's function is "emergent," having no meaning for the isolated enzyme molecule, the binding process, and the forces employed in it, can be given a wholly reductionist explanation in terms of the known physico-chemical properties of the molecules that participate in it. Consequently, this weak form of emergence poses no problems for even the most ardent reductionist.

"Weak emergence" shows up in a variety of ways. In describing a complex system we often find it convenient to introduce new theoretical terms, like inertial mass in mechanics, or voltage in the theory of circuits, for quantities that are not directly observable but are defined by relations among the observables.[2] We can often use such terms to avoid reference to details of the component subsystems, referring only to their aggregate properties.

Ohm, for example, established his law of electrical resistance by constructing a circuit containing a battery that drove current through a wire, and an ammeter that measured the magnetic force induced by the current. By changing the length of the wire, he altered the current. The equation relating the length of the wire (resistance) to the force registered by the ammeter (current) contained two constants, which were independent of the length of the wire but changed if he replaced the battery by another. These constants were labeled the *voltage* and *internal resistance* of the battery, which was otherwise unanalyzed and treated as a "black box." Voltage and internal resistance are not measured directly but are theoretical terms, inferred from the measured resistance and current with the aid of Ohm's Law.

Whereas the details of components can often be ignored while studying their interactions in the whole system, the short-run behavior of the individual subsystems can often be described in detail while ignoring the (slower) interactions among subsystems. In economics, we often study the interaction of closely related markets—for example, the markets for iron ore, pig iron, sheet steel and steel products—under the assumption that all other supply and demand relations remain constant. In the next chapter, we will discuss at length this near independence of hierarchical

2. H. A. Simon, "The Axiomatization of Physical Theories," *Philosophy of Science*, 37(1970), 16–26.

systems from the detail of their component subsystems, as well as the short-run independence of the subsystems from the slower movements of the total system.

By adopting this weak interpretation of emergence, we can adhere (and I will adhere) to reductionism in principle even though it is not easy (often not even computationally feasible) to infer rigorously the properties of the whole from knowledge of the properties of the parts. In this pragmatic way, we can build nearly independent theories for each successive level of complexity, but at the same time, build bridging theories that show how each higher level can be accounted for in terms of the elements and relations of the next level below.

This is, of course, the usual conception of the sciences as building upward from elementary particles, through atoms and molecules to cells, organs and organisms. The actual history, however, has unfolded, as often as not, in the opposite direction—from top down. We have already observed, in chapter 1, how we commonly hang our scientific theories from skyhooks.

Cybernetics and General Systems Theory

The period during and just after World War II saw the emergence of what Norbert Wiener dubbed "cybernetics": a combination of servomechanism theory (feedback control systems), information theory, and modern stored-program computers, all of which afford bold new insights into complexity. Information theory explains organized complexity in terms of the reduction of entropy (disorder) that is achieved when systems (organisms, for example) absorb energy from external sources and convert it to pattern or structure. In information theory, energy, information, and pattern all correspond to negative entropy.

Feedback control shows how a system can work toward goals and adapt to a changing environment,[3] thereby removing the mystery from teleology. What is required is ability to recognize the goal, to detect differences between the current situation and the goal, and actions that can reduce such differences: precisely the capabilities embodied in a system like the General Problem Solver. Soon this insight was being applied to

---

3. A. Rosenblueth, N. Wiener and J. Bigelow, "Behavior, Purpose and Teleology," *Philosophy of Science*, 10(1943), 18–24.

constructing small robots that could maneuver around a room autonomously.[4] As computers became available, systems could be built at levels of complexity that had never before been contemplated; and by virtue of their capability for interpreting and executing their own internally stored programs, computers initiated the study of artificial intelligence.

These developments encouraged both the study of complex systems, especially adaptive goal-oriented systems, "as wholes," and simultaneously, the reductive explanation of system properties in terms of mechanisms. Holism was brought into confrontation with reductionism in a way that had never been possible before, and that confrontation continues today in philosophical discussion of artificial systems.

During these postwar years, a number of proposals were advanced for the development of "general systems theory," that, abstracting from the special properties of physical, biological, or social systems, would apply to all of them.[5] We might well feel that, while the goal is laudable, systems of these diverse kinds could hardly be expected to have any nontrivial properties in common. Metaphor and analogy can be helpful or they can be misleading. All depends on whether the similarities the metaphor captures are significant or superficial.

If a general systems theory is too ambitious a goal, it might still not be vain to search for common properties among broad *classes* of complex systems. The ideas that go by the name of cybernetics constitute, if not a theory, at least a point of view that has proved fruitful over a wide range of applications.[6] It has been highly useful to look at the behavior of adaptive systems in terms of feedback and homeostasis and to apply to these concepts the theory of selective information.[7] The concepts of feedback

4. W. Grey Walter, "An Imitation of Life," *Scientific American, 182*(5) (1950):42.

5. See especially the yearbooks of the Society for General Systems Research. Prominent exponents of general systems theory were L. von Bertalanffy, K. Boulding, R. W. Gerard and, still active in this endeavor, J. G. Miller.

6. N. Wiener, *Cybernetics* (New York: Wiley, 1948). For an imaginative forerunner, see A. J. Lotka, *Elements of Mathematical Biology* (New York: Dover Publications, 1951), first published in 1924 as *Elements of Physical Biology.*

7. C. Shannon and W. Weaver, *The Mathematical Theory of Communication* (Urbana: University of Illinois Press, 1949); W. R. Ashby, *Design for a Brain* (New York: Wiley, 1952).

and information provide a frame of reference for viewing a wide range of situations, just as do the ideas of evolution, of relativism, of axiomatic method, and of operationalism.

The principal contribution of this second wave of inquiry into complexity lay precisely in the more specific concepts it brought to attention rather than in the broad idea of a general systems theory. This view is illustrated in the next chapter, which focuses on the properties of those particular complex systems that are hierarchical in structure, and draws out the consequences for system behavior of the strong assumption of hierarchy (or near-decomposability, as I shall call it).

Current Interest in Complexity

The current, third, burst of interest in complexity shares many of the characteristics of the second. Much of the motivation for it is the growing need to understand and cope with some of the world's large-scale systems—the environment, for one, the world-wide society that our species has created, for another, and organisms, for a third. But this motivation could not, by itself, tie attention to complexity for very long if novel ways of thinking about it were not also provided. Going beyond the tools and concepts that appeared in the second wave, other new ideas have emerged, together with relevant mathematics and computational algorithms. The ideas have such labels as "catastrophe," "chaos," "genetic algorithms," and "cellular automata."

As always, the labels have some tendency to assume a life of their own. The foreboding tone of "catastrophe" and "chaos" says something about the age of anxiety in which these concepts were named. Their value as concepts, however, depends not on the rhetoric they evoke, but on their power to produce concrete answers to questions of complexity. For the particular concepts listed above, much of the verdict is not yet in. I want to comment briefly on each of them, for they are both alternatives and complements to the approach to hierarchical complexity that I will develop in the next chapter.

Catastrophe Theory

Catastrophe theory appeared on the scene around 1968,[8] made an audible splash, and nearly faded from public sight within a few years. It is a solid body of mathematics dealing with the classification of nonlinear dynamic systems according to their modes of behavior. Catastrophic events occur in a special kind of system. Such a system can assume two (or more) distinct steady states (static equilibria, for example, or periodic cycles); but when the system is in one of these states, a moderate change in a system parameter may cause it to shift suddenly to the other—or into an unstable state in which variables increase without limit. The mathematician R. Thom constructed a topological classification of two-variable and three-variable systems according to the kinds of catastrophes they could or couldn't experience.

It is not hard to think of natural systems that exhibit behavior of this kind—stable behavior followed by a sudden shift to disequilibrium or to another, quite different, equilibrium. A commonly cited example is the threatened dog that either suddenly moves to the attack—or panics and flees. More complex examples have been studied: for instance, a budworm population infesting a spruce forest. The rapidly reproducing budworms quickly reach an equilibrium of maximum density; but the slow continuing growth of the spruce forest gradually alters the limit on the budworm population until, when a critical forest density is exceeded, the population explodes.[9] One can conjure up models of human revolutions embodying similar mechanisms.

In the circumstances that create it, the catastrophe mechanism is effective and the metaphor evocative, but in practice, only a limited number of situations have been found where it leads to any further analysis. Most of the initial applications that struck public fancy (like the attacking/fleeing dog) were after-the-fact explanations of phenomenon that were

---

8. See R. Thom, *An Outline of a General Theory of Models* (Reading, MA: Benjamin, 1975).

9. For an account of the spruce/budworm model, see T. F. H. Allen and T. B. Starr, *Ecology: Perspectives for Ecological Complexity* (Chicago, IL: University of Chicago Press, 1982), and references cited there. In the next chapter we will see that this same example can be described as a nearly decomposable system.

already familiar. For this reason, catastrophe theory is much less prominent in the public eye and in the literature of complexity today than it was twenty-five years ago.

Complexity and Chaos

The theory of chaos also represents solid mathematics, which in this case has a long history reaching back to Poincaré.[10] Chaotic systems are deterministic dynamic systems that, if their initial conditions are disturbed even infinitesimally, may alter their paths radically. Thus, although they are deterministic, their detailed behavior over time is unpredictable, for small perturbations cause large changes in path. Chaotic systems were sufficiently intractable to mathematical treatment that, although the subject was kept alive by a few French mathematicians working in the tradition of Poincaré, only modest progress was made with them until well beyond the middle of this century. A major source of new progress has been the ability to use computers to display and explore their chaotic behavior.

Gradually, researchers in a number of sciences began to suspect that important phenomena they wished to understand were, in this technical sense, chaotic. One of the first was the meteorologist E. N. Lorenz, who started to explore in the early 1960s the possibility that weather was a chaotic phenomenon—the possibility that the butterfly in Singapore, by flapping its wings, could cause a thunderstorm in New York. Soon, fluid turbulence in general was being discussed in terms of chaos; and the possible inculpation of chaos in the complex behavior of a wide range of physical and biological systems was being studied. Solid experimental evidence that specific physical systems do, in fact, behave chaotically began to appear in the late 1970s.[11]

The growth in attention to chaos must be viewed against the background of our general understanding of dynamic systems. For a long time we have had a quite general theory of systems of *linear* differential equa-

10. H. Poincaré, *Les Methodes Nouvelle de la Méchanique Céleste.* (Paris: Gauthier-Villars, 1892).

11. An excellent selection of the literature of chaos, both mathematical and experimental, up to the middle 1980s can be found in P. Cvitanović (ed.), *Universality in Chaos* (Bristol: Adam Hilger, 1986).

tions and their solution in closed form. With systems of *nonlinear* equations, matters were less satisfactory. Under particular simple boundary conditions, solutions were known for a number of important systems of nonlinear partial differential equations that capture the laws of various kinds of wave motion. But beyond these special cases, knowledge was limited to methods for analyzing local behavior qualitatively—its stability or instability—in order to divide the space of achievable states into discrete regions. In each such region, specific kinds of behavior (e.g., movement to equilibrium, escape from unstable equilibrium, steady-state motion in limit cycles) would occur.[12]

This was the bread-and-butter content of the standard textbook treatments of nonlinear analysis, and beyond these qualitative generalizations, complex nonlinear systems had to be studied mainly by numerical simulation with computers. Most of the large computers and super-computers of the past half century have been kept busy simulating numerically the behavior of the systems of partial differential equations that describe the dynamics of airplanes, atomic piles, the atmosphere, and turbulent systems generally. As chaotic systems were not typically discussed in the textbooks, the then-current theory of nonlinear systems provided little help in treating such phenomena as turbulence except at an aggregate and very approximate level.

Under these circumstances, new computer-generated discoveries about chaos in the late 1970s and early 1980s created enormous interest and excitement in a variety of fields where phenomena were already suspected of being chaotic, and hence could perhaps be understood better with the new theory. Numerical computations on simple nonlinear systems revealed unsuspected invariants ("universal numbers") that predicted, for broad classes of such systems, at what point they would change from orderly to chaotic behavior.[13] Until high-speed computers were available to reveal them, such regularities were invisible.

12. A standard source is A. A. Andronov, E. A. Leontovich, I. I. Gordon and A. G. Maier, *Qualitative Theory of Second-Order Dynamic Systems* (Wiley, NY: 1973).

13. M. J. Feigenbaum, "Universal Behavior in Nonlinear Systems," *Los Alamos Science*, 1(1980):4–27. This and other "classic" papers on chaos from the 1970s and 1980s are reprinted in P. Cvitanović, ed., *op. cit.*

Deep understanding has now been achieved of many aspects of chaos, but to say that we "understand" does not imply that we can predict. Chaos led to the recognition of a new, generalized, notion of equilibrium—the so-called "strange attractor." In classical nonlinear theory a system might come to a stable equilibrium, or it might oscillate permanently in a limit cycle, like the orbit of a planet. A chaotic system, however, might also enter a region of its state space, the strange attractor, in which it would remain permanently.

Within the strange attractor, motion would not cease, nor would it be predictable, but although deterministic, would appear to be random. That is, slightly different directions of entrance into the strange attractor, or slight perturbations when in it, would lead the system into quite different paths. A billiard ball aimed exactly at a 45° angle across a square "ideal" billiard table, will reflect off successive sides and, returning to the starting point, repeat its rectangular path indefinitely. But if you decrease or increase the 45° angle by an epsilon, the ball will never return to the starting point but will pursue a path that will eventually take it as close as you please to any spot on the table. The table's entire surface has become the strange attractor for the chaotic behavior and almost equal but different initial angles will produce continually diverging paths.

The theory of chaos has perhaps not maintained the hectic pace of development it experienced from the early 1960s to the late 1980s, but during this period it established itself as an essential conceptual framework and mathematical tool for the study of a class of systems that have major real-world importance in a number of scientific domains. The mechanisms of chaos are more general, but also of wider application, than those of catastrophe theory. Hence, we can expect chaos to continue to play a larger role than catastrophe in the continuing study of complex systems.

Rationality in a Catastrophic or Chaotic World

What implications do catastrophe and chaos have for the systems—economies, the human mind, and designed complex systems—that we have been discussing in the previous six chapters? Although there have been some attempts to discover chaos in economic time series, the results thus far have been inconclusive. I am aware of no clear demonstration of chaos in the brain, but there is increasing evidence that chaos plays a role, although still a rather unclear one, in the functioning of the normal and defective heart. Designers frequently construct systems (e.g., airplanes and ships) that produce, and cope successfully with, turbulence and perhaps other kinds of chaos.

On the basis of the evidence, we should suppose neither that all of the complex systems we encounter in the world are chaotic, nor that few of them are. Moreover, as the airplane example shows, the ominous term "chaotic" should not be read as "unmanageable." Turbulence is frequently present in hydraulic and aerodynamic situations and artifacts. In such situations, although the future is not predictable in any detail, it is manageable as an aggregate phenomenon. And the paths of tornadoes and hurricanes are notoriously unstable but stable enough in the short run that we can usually be warned and reach shelter before they hit us.

Since Newton, astronomers have been able to compute the motion of a system of two bodies that exercise mutual gravitational attraction on each other. With three or more bodies, they never obtained more than approximations to the motion, and indeed, there is now good reason to believe that, in general, gravitational systems of three or more bodies, including the solar system, are chaotic. But we have no reason to anticipate untoward consequences from that chaos—its presence simply implies that astronomers will be frustrated in their attempts to predict the exact positions of the planets in the rather long run—a perplexity as frustrating as, but perhaps less damaging than, the difficulties meteorologists experience in predicting the weather.

Finally, there has been substantial progress in devising feedback devices that "tame" chaos by restricting chaotic systems, moving within their strange attractors, to small neighborhoods having desired properties, so that the chaos becomes merely tolerable noise. Such devices provide an example, consonant with the discussion in earlier chapters, of the substitution of control for prediction.

Complexity and Evolution

Much current research on complex systems focuses upon the emergence of complexity, that is, system evolution. Two computational approaches to evolution that have attracted particular attention are the genetic algorithms first explored by Holland[14] and computer algorithms for cellular automata that simulate the multiplication and competition of organisms, playing the so-called "game of life."

Genetic Algorithms.   From an evolutionary standpoint, an organism can be represented by a list or vector of features (its genes). Evolution evaluates this vector in terms of fitness for survival. From generation to generation, the frequency distribution of features and their combinations over the members of a species change through sexual reproduction, crossover, inversion, and mutation. Natural selection causes features and combinations of features contributing to high fitness to multiply more rapidly than, and ultimately to replace, features and combinations conducive to low fitness.

By programming this abstraction on a modern computer we can build a computational model of the process of evolution. The simulation, in turn, can be used to study the relative rates at which fitness will grow under different assumptions about the model, including assumptions about rates of mutation and crossover. In the next chapter, we will consider the special case of evolution in hierarchical systems, which appears to be the kind of system that predominates in the natural world.

Cellular Automata and The Game of Life.   The computer is used not only to estimate the statistics of evolution but to carry out simulations, at an abstract level, of evolutionary processes. This research goes back, in fact, to the second eruption of interest in complexity, after World War II, when John von Neumann, building on some ideas of Stanislaw Ulam, defined abstractly (but did not implement) a system that was capable of reproducing itself. The idea was kept alive by Arthur Burks and others, but it was not until well into the current period of activity that Christopher Langton created a computer program that simulated a self-reproducing cellular automaton.[15] Computer programs can create

14. J. H. Holland, *Adaptation in Natural and Artificial Systems* (Ann Arbor: University of Michigan Press, 1975).

15. A. W. Burks (ed.), *Essays on Cellular Automata* (Champaign-Urbana: University of Illinois Press, 1970); C. G. Langton (ed.), *Artificial Life*. Santa Fe Insti-

symbolic objects of various kinds and apply rules for their replication or destruction as a function of their environments (which include other nearby objects). With appropriate selection of the system parameters, such simulations can provide vivid demonstrations of evolving self-reproducing systems. This line of exploration is still at a very early stage of development, is largely dependent on computer simulation, and lacks any large body of formal theory. It will be some time before we can assess its potential, but it has already presented us with a fundamental and exciting result: self-reproducing systems are a reality.

Conclusion

Complexity is more and more acknowledged to be a key characteristic of the world we live in and of the systems that cohabit our world. It is not new for science to attempt to understand complex systems: astronomers have been at it for millennia, and biologists, economists, psychologists, and others joined them some generations ago. What is new about the present activity is not the study of particular complex systems but the study of the phenomenon of complexity in its own right.

If, as appears to be the case, complexity (like systems science) is too general a subject to have much content, then particular classes of complex systems possessing strong properties that provide a fulcrum for theorizing and generalizing can serve as the foci of attention. More and more, this appears to be just what is happening, with chaos, genetic algorithms, cellular automata, catastrophe, and hierarchical systems serving as some of the currently visible focal points. In the next chapter we will examine the last-named of these more closely.

---

tute Studies in the Sciences of Complexity, Proceedings, vol. 6 (Redwood City, CA: Addison-Wesley, 1989); C. G. Langton, C. Taylor, J. D. Farmer and S. Rassmussen (eds.), *Artificial Life II*. Santa Fe Institute Studies in the Sciences of Complexity, Proceedings, vol. 10 (Redwood City, CA: Addison-Wesley, 1992).

8

The Architecture of Complexity: Hierarchic Systems

In this chapter I should like to report on some things we have been learning about particular kinds of complex systems encountered in various sciences. The developments I shall discuss arose in the context of specific phenomena, but the theoretical formulations themselves make little reference to details of structure. Instead they refer primarily to the complexity of the systems under view without specifying the exact content of that complexity. Because of their abstractness, the theories may have relevance—application would be too strong a term—to many kinds of complex systems observed in the social, biological, and physical sciences.

In recounting these developments, I shall avoid technical detail, which can generally be found elsewhere. I shall describe each theory in the particular context in which it arose. Then I shall cite some examples of complex systems, from areas of science other than the initial application, to which the theoretical framework appears relevant. In doing so, I shall make reference to areas of knowledge where I am not expert—perhaps not even literate. The reader will have little difficulty, I am sure, in distinguishing instances based on idle fancy or sheer ignorance from instances that cast some light on the ways in which complexity exhibits itself wherever it is found in nature.

I shall not undertake a formal definition of "complex systems."[1] Roughly, by a complex system I mean one made up of a large number of

This chapter is a revision of a paper with the same title, reprinted with permission from *Proceedings of the American Philosophical Society, 106*(December 1962): 467–482.

1. W. Weaver, in "Science and Complexity," *American Scientist, 36*(1948):536, has distinguished two kinds of complexity, disorganized and organized. We shall be concerned primarily with organized complexity.

parts that have many interactions. As we saw in the last chapter, in such systems the whole is more than the sum of the parts in the weak but important pragmatic sense that, given the properties of the parts and the laws of their interaction, it is not a trivial matter to infer the properties of the whole.[2]

The four sections that follow discuss four aspects of complexity. The first offers some comments on the frequency with which complexity takes the form of hierarchy—the complex system being composed of subsystems that in turn have their own subsystems, and so on. The second section theorizes about the relation between the structure of a complex system and the time required for it to emerge through evolutionary processes; specifically it argues that hierarchic systems will evolve far more quickly than nonhierarchic systems of comparable size. The third section explores the dynamic properties of hierarchically organized systems and shows how they can be decomposed into subsystems in order to analyze their behavior. The fourth section examines the relation between complex systems and their descriptions.

Thus my central theme is that complexity frequently takes the form of hierarchy and that hierarchic systems have some common properties independent of their specific content. Hierarchy, I shall argue, is one of the central structural schemes that the architect of complexity uses.

Hierarchic Systems

By a *hierarchic system,* or hierarchy, I mean a system that is composed of interrelated subsystems, each of the latter being in turn hierarchic in structure until we reach some lowest level of elementary subsystem. In most systems in nature it is somewhat arbitrary as to where we leave off

2. See also John R. Platt, "Properties of Large Molecules that Go beyond the Properties of Their Chemical Sub-groups," *Journal of Theoretical Biology,* 1(1961):342–358. Since the reductionism-holism issue is a major *cause de guerre* between scientists and humanists, perhaps we might even hope that peace could be negotiated between the two cultures along the lines of the compromise just suggested. As I go along, I shall have a little to say about complexity in the arts as well as in the natural sciences. I must emphasize the pragmatism of my holism to distinguish it sharply from the position taken by W. M. Elsasser in *The Physical Foundation of Biology* (New York: Pergamon Press, 1958).

the partitioning and what subsystems we take as elementary. Physics makes much use of the concept of "elementary particle," although particles have a disconcerting tendency not to remain elementary very long. Only a couple of generations ago the atoms themselves were elementary particles; today to the nuclear physicist they are complex systems. For certain purposes of astronomy whole stars, or even galaxies, can be regarded as elementary subsystems. In one kind of biological research a cell may be treated as an elementary subsystem; in another, a protein molecule; in still another, an amino acid residue.

Just why a scientist has a right to treat as elementary a subsystem that is in fact exceedingly complex is one of the questions we shall take up. For the moment we shall accept the fact that scientists do this all the time and that, if they are careful scientists, they usually get away with it.

Etymologically the word "hierarchy" has had a narrower meaning than I am giving it here. The term has generally been used to refer to a complex system in which each of the subsystems is subordinated by an authority relation to the system it belongs to. More exactly, in a hierarchic formal organization each system consists of a "boss" and a set of subordinate subsystems. Each of the subsystems has a "boss" who is the immediate subordinate of the boss of the system. We shall want to consider systems in which the relations among subsystems are more complex than in the formal organizational hierarchy just described. We shall want to include systems in which there is no relation of subordination among subsystems. (In fact even in human organizations the formal hierarchy exists only on paper; the real flesh-and-blood organization has many interpart relations other than the lines of formal authority.) For lack of a better term I shall use "hierarchy" in the broader sense introduced in the previous paragraphs to refer to all complex systems analyzable into successive sets of subsystems and speak of "formal hierarchy" when I want to refer to the more specialized concept.[3]

3. The mathematical term "partitioning" will not do for what I call here a hierarchy; for the set of subsystems and the successive subsets in each of these define the partitioning, independent of any systems of relations among the subsets. By "hierarchy" I mean the partitioning in conjunction with the relations that hold among its parts.

Social Systems

I have already given an example of one kind of hierarchy that is frequently encountered in the social sciences—a formal organization. Business firms, governments, and universities all have a clearly visible parts-within-parts structure. But formal organizations are not the only, or even the most common, kind of social hierarchy. Almost all societies have elementary units called families, which may be grouped into villages or tribes, and these into larger groupings, and so on. If we make a chart of social interactions, of who talks to whom, the clusters of dense interaction in the chart will identify a rather well-defined hierarchic structure. The groupings in this structure may be defined operationally by some measure of frequency of interaction in this sociometric matrix.

Biological and Physical Systems

The hierarchical structure of biological systems is a familiar fact. Taking the cell as the building block, we find cells organized into tissues, tissues into organs, organs into systems. Within the cell are well-defined subsystems—for example, nucleus, cell membrane, microsomes, and mitochondria.

The hierarchic structure of many physical systems is equally clear-cut. I have already mentioned the two main series. At the microscopic level we have elementary particles, atoms, molecules, and macromolecules. At the macroscopic level we have satellite systems, planetary systems, galaxies. Matter is distributed throughout space in a strikingly nonuniform fashion. The most nearly random distributions we find, gases, are not random distributions of elementary particles but random distributions of complex systems, that is, molecules.

A considerable range of structural types is subsumed under the term "hierarchy" as I have defined it. By this definition a diamond is hierarchic, for it is a crystal structure of carbon atoms that can be further decomposed into protons, neutrons, and electrons. However, it is a very "flat" hierarchy, in which the number of first-order subsystems belonging to the crystal can be indefinitely large. A volume of molecular gas is a flat hierarchy in the same sense. In ordinary usage we tend to reserve the word "hierarchy" for a system that is divided into a *small or moderate number* of subsystems, each of which may be further subdivided. Hence we do not ordinarily think of or refer to a diamond or a gas as a hierarchic structure. Similarly a linear polymer is simply a chain, which may be very long, of identical subparts, the monomers. At the molecular level it is a very flat hierarchy.

In discussing formal organizations, the number of subordinates who report directly to a single boss is called his *span of control*. I shall speak analogously of the *span* of a system, by which I shall mean the number of subsystems into which it is partitioned. Thus a hierarchic system is flat at a given level if it has a wide span at that level. A diamond has a wide span at the crystal level but not at the next level down, the atomic level.

In most of our theory construction in the following sections we shall focus our attention on hierarchies of moderate span, but from time to time I shall comment on the extent to which the theories might or might not be expected to apply to very flat hierarchies.

There is one important difference between the physical and biological hierarchies, on the one hand, and social hierarchies, on the other. Most physical and biological hierarchies are described in spatial terms. We detect the organelles in a cell in the way we detect the raisins in a cake—they are "visibly" differentiated substructures localized spatially in the larger structure. On the other hand, we propose to identify social hierarchies not by observing who lives close to whom but by observing who interacts with whom. These two points of view can be reconciled by defining hierarchy in terms of intensity of interaction, but observing that in most biological and physical systems relatively intense interaction implies relative spatial propinquity. One of the interesting characteristics of nerve cells and telephone wires is that they permit very specific strong interactions at great distances. To the extent that interactions are channeled through specialized communications and transportation systems, spatial propinquity becomes less determinative of structure.

Symbolic Systems

One very important class of systems has been omitted from my examples thus far: systems of human symbolic production. A book is a hierarchy in the sense in which I am using that term. It is generally divided into chapters, the chapters into sections, the sections into paragraphs, the paragraphs into sentences, the sentences into clauses and phrases, the clauses and phrases into words. We may take the words as our elementary units, or further subdivide them, as the linguist often does, into smaller units. If the book is narrative in character, it may divide into "episodes" instead of sections, but divisions there will be.

The hierarchic structure of music, based on such units as movements, parts, themes, phrases, is well known. The hierarchic structure of products of the pictorial arts is more difficult to characterize, but I shall have something to say about it later.

The Evolution of Complex Systems

Let me introduce the topic of evolution with a parable. There once were two watchmakers, named Hora and Tempus, who manufactured very fine watches. Both of them were highly regarded, and the phones in their workshops rang frequently—new customers were constantly calling them. However, Hora prospered, while Tempus became poorer and poorer and finally lost his shop. What was the reason?

The watches the men made consisted of about 1,000 parts each. Tempus had so constructed his that if he had one partly assembled and had to put it down—to answer the phone, say—it immediately fell to pieces and had to be reassembled from the elements. The better the customers liked his watches, the more they phoned him and the more difficult it became for him to find enough uninterrupted time to finish a watch.

The watches that Hora made were no less complex than those of Tempus. But he had designed them so that he could put together subassemblies of about ten elements each. Ten of these subassemblies, again, could be put together into a larger subassembly; and a system of ten of the latter subassemblies constituted the whole watch. Hence, when Hora had to put down a partly assembled watch to answer the phone, he lost only a small part of his work, and he assembled his watches in only a fraction of the man-hours it took Tempus.

It is rather easy to make a quantitative analysis of the relative difficulty of the tasks of Tempus and Hora: suppose the probability that an interruption will occur, while a part is being added to an incomplete assembly, is p. Then the probability that Tempus can complete a watch he has started without interruption is $(1 - p)^{1000}$—a very small number unless p

is 0.001 or less. Each interruption will cost on the average the time to assemble $1/p$ parts (the expected number assembled before interruption). On the other hand, Hora has to complete 111 subassemblies of ten parts each. The probability that he will not be interrupted while completing any one of these is $(1 - p)^{10}$, and each interruption will cost only about the time required to assemble five parts.[4]

Now if p is about 0.01—that is, there is one chance in a hundred that either watchmaker will be interrupted while adding any one part to an assembly—then a straightforward calculation shows that it will take Tempus on the average about four thousand times as long to assemble a watch as Hora.

We arrive at the estimate as follows:

1. Hora must make 111 times as many complete assemblies per watch as Tempus; but

2. Tempus will lose on the average 20 times as much work for each interrupted assembly as Hora (100 parts, on the average, as against 5); and

3. Tempus will complete an assembly only 44 times per million attempts $(0.99^{1000} = 44 \times 10^{-6})$, while Hora will complete nine out of ten $(0.99^{10} = 9 \times 10^{-1})$. Hence Tempus will have to make 20,000 as many attempts per completed assembly as Hora. $(9 \times 10^{-1})/(44 \times 10^{-6}) = 2 \times 10^4$. Multiplying these three ratios, we get

4. The speculations on speed of evolution were first suggested by H. Jacobson's application of information theory to estimating the time required for biological evolution. See his paper "Information, Reproduction, and the Origin of Life," in *American Scientist*, 43(January 1955):119–127. From thermodynamic considerations it is possible to estimate the amount of increase in entropy that occurs when a complex system decomposes into its elements. (See for example, R. B. Setlow and E. C. Pollard, *Molecular Biophysics* (Reading, Mass.: Addison-Wesley, 1962), pp. 63–65, and references cited there.) But entropy is the logarithm of a probability; hence information, the negative of entropy, can be interpreted as the logarithm of the reciprocal of the probability—the "improbability," so to speak. The essential idea in Jacobson's model is that the expected time required for the system to reach a particular state is inversely proportional to the probability of the state—hence it increases exponentially with the amount of information (negentropy) of the state.

Following this line of argument, but not introducing the notion of levels and stable subassemblies, Jacobson arrived at estimates of the time required for evolution so large as to make the event rather improbable. Our analysis, carried through in the same way, but with attention to the stable intermediate forms, produces very much smaller estimates.

$1/111 \times 100/5 \times 0.99^{10}/0.99^{1000}$
$= 1/111 \times 20 \times 20,000 \sim 4,000.$

Biological Evolution

What lessons can we draw from our parable for biological evolution? Let us interpret a partially completed subassembly of k elementary parts as the coexistence of k parts in a small volume—ignoring their relative orientations. The model assumes that parts are entering the volume at a constant rate but that there is a constant probability, p, that the part will be dispersed before another is added, unless the assembly reaches a stable state. These assumptions are not particularly realistic. They undoubtedly underestimate the decrease in probability of achieving the assembly with increase in the size of the assembly. Hence the assumptions understate— probably by a large factor—the relative advantage of a hierarchic structure.

Although we cannot therefore take the numerical estimate seriously, the lesson for biological evolution is quite clear and direct. The time required for the evolution of a complex form from simple elements depends critically on the numbers and distribution of potential intermediate stable forms. In particular, if there exists a hierarchy of potential stable "subassemblies," with about the same span, s, at each level of the hierarchy, then the time required for a subassembly can be expected to be about the same at each level—that is, proportional to $1/(1-p)^s$. The time required for the assembly of a system of n elements will be proportional to $\log_s n$, that is, to the number of levels in the system. One would say—with more illustrative than literal intent—that the time required for the evolution of multi-celled organisms from single-celled organisms might be of the same order of magnitude as the time required for the evolution of single-celled organisms from macromolecules. The same argument could be applied to the evolution of proteins from amino acids, of molecules from atoms, of atoms from elementary particles.

A whole host of objections to this oversimplified scheme will occur, I am sure, to every working biologist, chemist, and physicist. Before turning to matters I know more about, I shall lay at rest four of these problems, leaving the remainder to the attention of the specialists.

First, in spite of the overtones of the watchmaker parable, the theory assumes no teleological mechanism. The complex forms can arise from the simple ones by purely random processes. (I shall propose another model in a moment that shows this clearly.) Direction is provided to the scheme by the stability of the complex forms, once these come into existence. But this is nothing more than survival of the fittest—that is, of the stable.

Second, not all large systems appear hierarchical. For example, most polymers—such as nylon—are simply linear chains of large numbers of identical components, the monomers. However, for present purposes we can simply regard such a structure as a hierarchy with a span of one—the limiting case; for a chain of any length represents a state of relative equilibrium.[5]

Third, the evolutionary process does not violate the second law of thermodynamics. The evolution of complex systems from simple elements implies nothing, one way or the other, about the change in entropy of the entire system. If the process absorbs free energy, the complex system will have a smaller entropy than the elements; if it releases free energy, the opposite will be true. The former alternative is the one that holds for most biological systems, and the net inflow of free energy has to be supplied from the sun or some other source if the second law of thermodynamics is not to be violated. For the evolutionary process we are describing, the equilibria of the intermediate states need have only local and not global stability, and they may be stable only in the steady state—that is, as long as there is an external source of free energy that may be drawn upon.[6]

5. There is a well-developed theory of polymer size, based on models of random assembly. See, for example, P. J. Flory, *Principles of Polymer Chemistry* (Ithaca: Cornell University Press, 1953), chapter 8. Since *all* subassemblies in the polymerization theory are stable, limitation of molecular growth depends on "poisoning" of terminal groups by impurities or formation of cycles rather than upon disruption of partially formed chains.

6. This point has been made many times before, but it cannot be emphasized too strongly. For further discussion, see Setlow and Pollard, *Molecular Biophysics,* pp. 49–64; E. Schrödinger, *What Is Life?* (Cambridge: Cambridge University Press, 1945); and H. Linschitz, "The Information Content of a Bacterial Cell," in H. Quastler (ed.), *Information Theory in Biology* (Urbana: University of Illinois Press, 1953), pp. 251–262.

Because organisms are not energetically closed systems, there is no way to deduce the direction, much less the rate, of evolution from classical thermodynamic considerations. All estimates indicate that the amount of entropy, measured in physical units, involved in the formation of a one-celled biological organism is trivially small—about -10^{-11} cal/degree.[7] The "improbability" of evolution has nothing to do with this quantity of entropy, which is produced by every bacterial cell every generation. The irrelevance of quantity of information, in this sense, to speed of evolution can also be seen from the fact that exactly as much information is required to "copy" a cell through the reproductive process as to produce the first cell through evolution.

The fact of the existence of stable intermediate forms exercises a powerful effect on the evolution of complex forms that may be likened to the dramatic effect of catalysts upon reaction rates and steady-state distribution of reaction products in open systems.[8] In neither case does the entropy change provide us with a guide to system behavior.

Evolution of Multi-Cellular Organisms

We must consider a fourth objection to the watchmaker metaphor. However convincing a model the metaphor may provide for the evolution of atomic and molecular systems, and even uni-cellular organisms, it does not appear to fit the history of multi-cellular organisms. The metaphor assumes that complex systems are formed by combining sets of simpler systems, but this is not the way in which multi-cellular organisms have evolved. Although bacteria may, in fact, have been produced by a merging of mitochondria with the cells they inhabited, multi-cellular organisms have evolved through multiplication and specialization of the cells of a single system, rather than through the merging of previously independent subsystems.

Lest we dismiss the metaphor too quickly, however, we should observe that systems that evolve through specialization acquire the same kind

7. See Linschitz, "The Information Content." This quantity, 10^{-11} cal/degree, corresponds to about 10^{13} bits of information.

8. See H. Kacser, "Some Physico-chemical Aspects of Biological Organization," appendix, pp. 191–249, in C. H. Waddington, *The Strategy of the Genes* (London: George Allen and Unwin, 1957).

of boxes-within-boxes structure (e.g., a digestive system consisting of mouth, larynx, esophagus, stomach, small and large intestines, colon; or a circulatory system consisting of a heart, arteries, veins, and capillaries) as is acquired by systems that evolve by assembly of simpler systems. The next main section of this chapter deals with nearly decomposable systems. It proposes that it is not assembly from components, per se, but hierarchic structure produced *either* by assembly or specialization, that provides the potential for rapid evolution.

The claim is that the potential for rapid evolution exists in any complex system that consists of a set of stable subsystems, each operating nearly independently of the detailed processes going on within the other subsystems, hence influenced mainly by the net inputs and outputs of the other subsystems. If the near-decomposability condition is met, the efficiency of one component (hence its contribution to the organism's fitness) does not depend on the detailed structure of other components.

Before examining this claim in detail, however, I should like to discuss briefly some non-biological applications of the watchmaker metaphor to illustrate the important advantages that hierarchic systems enjoy in other circumstances.

Problem Solving as Natural Selection

Hierarchy, as well as processes akin to natural selection, appear in human problem solving, a domain that has no obvious connection with biological evolution. Consider, for example, the task of discovering the proof for a difficult theorem. The process can be—and often has been—described as a search through a maze. Starting with the axioms and previously proved theorems, various transformations allowed by the rules of the mathematical systems are attempted, to obtain new expressions. These are modified in turn until, with persistence and good fortune, a sequence or path of transformations is discovered that leads to the goal.

The process ordinarily involves much trial and error. Various paths are tried; some are abandoned, others are pushed further. Before a solution is found, many paths of the maze may be explored. The more difficult and novel the problem, the greater is likely to be the amount of trial and error required to find a solution. At the same time the trial and error is not completely random or blind; it is in fact rather highly selective. The new expressions that are obtained by transforming given ones are examined to see whether they represent progress toward the goal. Indications of progress spur further search in the same direction; lack of progress signals the abandonment of a line of search. Problem solving requires *selective* trial and error.[9]

A little reflection reveals that cues signaling progress play the same role in the problem-solving process that stable intermediate forms play in the biological evolutionary process. In fact we can take over the watchmaker parable and apply it also to problem solving. In problem solving, a partial result that represents recognizable progress toward the goal plays the role of stable subassembly.

Suppose that the task is to open a safe whose lock has 10 dials, each with 100 possible settings, numbered from 0 to 99. How long will it take to open the safe by a blind trial-and-error search for the correct setting? Since there are 100^{10} possible settings, we may expect to examine about one half of these, on the average, before finding the correct one—that is, 50 billion billion settings. Suppose, however, that the safe is defective, so that a click can be heard when any one dial is turned to the correct setting. Now each dial can be adjusted independently and does not need to be touched again while the others are being set. The total number of settings that have to be tried is only 10×50, or 500. The task of opening the safe has been altered, by the cues the clicks provide, from a practically impossible one to a trivial one.[10]

9. See A. Newell, J. C. Shaw, and H. A. Simon, "Empirical Explorations of the Logic Theory Machine," *Proceedings of the 1957 Western Joint Computer Conference,* February 1957 (New York: Institute of Radio Engineers); "Chess-Playing Programs and the Problem of Complexity," *IBM Journal of Research and Development,* 2(October 1958):320–335; and for a similar view of problem solving, W. R. Ashby, "Design for an Intelligence Amplifier," pp. 215–233 in C. E. Shannon and J. McCarthy, *Automata Studies* (Princeton: Princeton University Press, 1956).

10. The clicking safe example was supplied by D. P. Simon. Ashby, "Design for an Intelligence Amplifier," p. 230, has called the selectivity involved in situations of this kind "selection by components." The even greater reduction in time produced by hierarchization in the clicking safe example, as compared with the watchmaker's metaphor, is due to the fact that a random *search* for the correct combination is involved in the former case, while in the latter the parts come together in the right order. It is not clear which of these metaphors provides the better model for biological evolution, but we may be sure that the watchmaker's

A considerable amount has been learned in the past thirty years about the nature of the mazes that represent common human problem-solving tasks—proving theorems, solving puzzles, playing chess, making investments, balancing assembly lines, to mention a few. All that we have learned about these mazes points to the same conclusion: that human problem solving, from the most blundering to the most insightful, involves nothing more than varying mixtures of trial and error and selectivity. The selectivity derives from various rules of thumb, or heuristics, that suggest which paths should be tried first and which leads are promising. We do not need to postulate processes more sophisticated than those involved in organic evolution to explain how enormous problem mazes are cut down to quite reasonable size (see also chapters 3 and 4).[11]

The Sources of Selectivity

When we examine the sources from which the problem-solving system, or the evolving system, as the case may be, derives its selectivity, we discover that selectivity can always be equated with some kind of feedback of information from the environment.

Let us consider the case of problem solving first. There are two basic kinds of selectivity. One we have already noted: various paths are tried out, the consequences of following them are noted, and this information is used to guide further search. In the same way in organic evolution various complexes come into being, at least evanescently, and those that are stable provide new building blocks for further construction. It is this information about stable configurations, and not free energy or negentropy from the sun, that guides the process of evolution and provides the selectivity that is essential to account for its rapidity.

The second source of selectivity in problem solving is previous experience. We see this particularly clearly when the problem to be solved is

---

metaphor gives an exceedingly conservative estimate of the savings due to hierarchization. The safe may give an excessively high estimate because it assumes all possible arrangements of the elements to be equally probable. For an application of a variant of the watchmaker and the clicking safe arguments to structure at the molecular level, see J. D. Watson, *Molecular Biology of the Gene*, 3rd ed. (Menlo Park, CA: W. A. Benjamin, 1976), pp. 107–108.

11. A. Newell and H. A. Simon, "Computer Simulation of Human Thinking," *Science, 134*(December 22, 1961):2011–2017.

similar to one that has been solved before. Then, by simply trying again the paths that led to the earlier solution, or their analogues, trial-and-error search is greatly reduced or altogether eliminated.

What corresponds to this latter kind of information in organic evolution? The closest analogue is reproduction. Once we reach the level of self-reproducing systems, a complex system, when it has once been achieved, can be multiplied indefinitely. Reproduction in fact allows the inheritance of acquired characteristics, but at the level of genetic material, of course; that is, only characteristics acquired by the genes can be inherited. We shall return to the topic of reproduction in the final section of this essay.

On Empires and Empire Building

We have not exhausted the categories of complex systems to which the watchmaker argument can reasonably be applied. Philip assembled his Macedonian empire and gave it to his son, to be later combined with the Persian subassembly and others into Alexander's greater system. On Alexander's death his empire did not crumble to dust but fragmented into some of the major subsystems that had composed it.

The watchmaker argument implies that if one would be Alexander, one should be born into a world where large stable political systems already exist. Where this condition was not fulfilled, as on the Scythian and Indian frontiers, Alexander found empire building a slippery business. So too, T. E. Lawrence's organizing of the Arabian revolt against the Turks was limited by the character of his largest stable building blocks, the separate, suspicious desert tribes.

The profession of history places a greater value upon the validated particular fact than upon tendentious generalization. I shall not elaborate upon my fancy therefore but shall leave it to historians to decide whether anything can be learned for the interpretation of history from an abstract theory of hierarchic complex systems.

Conclusion: The Evolutionary Explanation of Hierarchy

We have shown thus far that complex systems will evolve from simple systems much more rapidly if there are stable intermediate forms than if there are not. The resulting complex forms in the former case will be hierarchic. We have only to turn the argument around to explain the observed predominance of hierarchies among the complex systems nature presents to us. Among possible complex forms, hierarchies are the ones that have the time to evolve. The hypothesis that complexity will be hierarchic makes no distinction among very flat hierarchies, like crystals and tissues and polymers, and the intermediate forms. Indeed in the complex systems we encounter in nature examples of both forms are prominent. A more complete theory than the one we have developed here would presumably have something to say about the determinants of width of span in these systems.

Nearly Decomposable Systems

In hierarchic systems we can distinguish between the interactions *among* subsystems, on the one hand, and the interactions *within* subsystems—that is, among the parts of those subsystems—on the other. The interactions at the different levels may be, and often will be, of different orders of magnitude. In a formal organization there will generally be more interaction, on the average, between two employees who are members of the same department than between two employees from different departments. In organic substances intermolecular forces will generally be weaker than molecular forces, and molecular forces weaker than nuclear forces.

In a rare gas the intermolecular forces will be negligible compared to those binding the molecules—we can treat the individual particles for many purposes as if they were independent of each other. We can describe such a system as *decomposable* into the subsystems comprised of the individual particles. As the gas becomes denser, molecular interactions become more significant. But over some range we can treat the decomposable case as a limit and as a first approximation. We can use a theory of perfect gases, for example, to describe approximately the behavior of actual gases if they are not too dense. As a second approximation we may move to a theory of *nearly decomposable* systems, in which the interactions among the subsystems are weak but not negligible.

At least some kinds of hierarchic systems can be approximated successfully as nearly decomposable systems. The main theoretical findings from the approach can be summed up in two propositions: (1) in a nearly decomposable system the short-run behavior of each of the component subsystems is approximately independent of the short-run behavior of the other components; (2) in the long run the behavior of any one of the components depends in only an aggregate way on the behavior of the other components.

Let me provide a very concrete simple example of a nearly decomposable system.[12] Consider a building whose outside walls provide perfect thermal insulation from the environment. We shall take these walls as the boundary of our system. The building is divided into a large number of rooms, the walls between them being good, but not perfect, insulators. The walls between rooms are the boundaries of our major subsystems. Each room is divided by partitions into a number of cubicles, but the partitions are poor insulators. A thermometer hangs in each cubicle. Suppose that at the time of our first observation of the system there is a wide variation in temperature from cubicle to cubicle and from room to room—the various cubicles within the building are in a state of thermal disequilibrium. When we take new temperature readings several hours later, what shall we find? There will be very little variation in temperature among the cubicles within each single room, but there may still be large temperature variations *among* rooms. When we take readings again several days later, we find an almost uniform temperature throughout the building; the temperature differences among rooms have virtually disappeared.

We can describe the process of equilibrium formally by setting up the usual equations of heat flow. The equations can be represented by the matrix of their coefficients, r_{ij}, where r_{ij} is the rate at which heat flows from the ith cubicle to the jth cubicle per degree difference in their tem-

12. This discussion of near decomposability is based upon H. A. Simon and A. Ando, "Aggregation of Variables in Dynamic Systems," *Econometrica, 29* (April 1961):111–138. The example is drawn from the same source, pp. 117–118. For subsequent development and applications of the theory see P. J. Courtois, *Decomposability: Queueing and Computer System Applications* (New York, NY: Academic Press, 1977); Y. Iwasaki and H. A. Simon, "Causality and Model Abstraction," *Artificial Intelligence,* 67(1994):143–194; and D. F. Rogers and R. D. Plante, "Estimating Equilibrium Probabilities for Band Diagonal Markov Chains Using Aggregation and Disaggregation Techniques," *Computers in Operations Research,* 20(1993):857–877.

	A1	A2	A3	B1	B2	C1	C2	C3
A1	—	100	—	2	—	—	—	—
A2	100	—	100	1	1	—	—	—
A3	—	100	—	—	2	—	—	—
B1	2	1	—	—	100	2	1	—
B2	—	1	2	100	—	—	1	2
C1	—	—	—	2	—	—	100	—
C2	—	—	—	1	1	100	—	100
C3	—	—	—	—	2	—	100	—

Figure 7
A hypothetical nearly decomposable system. In terms of the heat-exchange example of the text. A1, A2, and A3 may be interpreted as cubicles in one room, B1 and B2 as cubicles in a second room, and C1, C2, and C3 as cubicles in a third. The matrix entries then are the heat diffusion coefficients between cubicles:

A1	B1	C1
A2		C2
A3	B2	C3

peratures. If cubicles i and j do not have a common wall, r_{ij} will be zero. If cubicles i and j have a common wall and are in the same room, r_{ij} will be large. If cubicles i and j are separated by the wall of a room, r_{ij} will be nonzero but small. Hence, by grouping together all the cubicles that are in the same room, we can arrange the matrix of coefficients so that all its large elements lie inside a string of square submatrices along the main diagonal. All the elements outside these diagonal squares will be either zero or small (see figure 7). We may take some small number, ε, as the upper bound of the extradiagonal elements. We shall call a matrix having these properties a *nearly decomposable matrix.*

Now it has been proved that a dynamic system that can be described by a nearly decomposable matrix has the properties, stated earlier, of a nearly decomposable system. In our simple example of heat flow this means that in the short run each room will reach an equilibrium temperature (an average of the initial temperatures of its offices) nearly independently of the others and that each room will remain approximately in a state of equilibrium over the longer period during which an over-all temperature equilibrium is being established throughout the building.

After the intra-room short-run equilibria have been reached, a single thermometer in each room will be adequate to describe the dynamic behavior of the entire system—separate thermometers in each cubicle will be superfluous.

Near Decomposability of Social Systems

As a glance at figure 7 shows, near decomposability is a rather strong property for a matrix to possess, and the matrices that have this property will describe very special dynamic systems—vanishingly few systems out of all those that are thinkable. How few they will be depends of course on how good an approximation we insist upon. If we demand that epsilon be very small, correspondingly few dynamic systems will fit the definition. But we have already seen that in the natural world nearly decomposable systems are far from rare. On the contrary, systems in which each variable is linked with almost equal strength with almost all other parts of the system are far rarer and less typical.

In economic dynamics the main variables are the prices and quantities of commodities. It is empirically true that the price of any given commodity and the rate at which it is exchanged depend to a significant extent only on the prices and quantities of a few other commodities, together with a few other aggregate magnitudes, like the average price level or some over-all measure of economic activity. The large linkage coefficients are associated in general with the main flows of raw materials and semifinished products within and between industries. An input-output matrix of the economy, giving the magnitudes of these flows, reveals the nearly decomposable structure of the system—with one qualification. There is a consumption subsystem of the economy that is linked strongly to variables in most of the other subsystems. Hence we have to modify our notions of decomposability slightly to accommodate the special role of the consumption subsystem in our analysis of the dynamic behavior of the economy.

In the dynamics of social systems, where members of a system communicate with and influence other members, near decomposability is generally very prominent. This is most obvious in formal organizations, where the formal authority relation connects each member of the organization with one immediate superior and with a small number of subordinates.

Of course many communications in organizations follow other channels than the lines of formal authority. But most of these channels lead from any particular individual to a very limited number of his superiors, subordinates, and associates. Hence departmental boundaries play very much the same role as the walls in our heat example.

Physicochemical Systems

In the complex systems familiar in biological chemistry, a similar structure is clearly visible. Take the atomic nuclei in such a system as the elementary parts of the system, and construct a matrix of bond strengths between elements. There will be matrix elements of quite different orders of magnitude. The largest will generally correspond to the covalent bonds, the next to the ionic bonds, the third group to hydrogen bonds, still smaller linkages to van der Waals forces.[13] If we select an epsilon just a little smaller than the magnitude of a covalent bond, the system will decompose into subsystems—the constituent molecules. The smaller linkages will correspond to the intermolecular bonds.

It is well known that high-energy, high-frequency vibrations are associated with the smaller physical subsystems and low-frequency vibrations with the larger systems into which the subsystems are assembled. For example, the radiation frequencies associated with molecular vibrations are much lower than those associated with the vibrations of the planetary electrons of the atoms; the latter in turn are lower than those associated with nuclear processes.[14] Molecular systems are nearly decomposable systems, with the short-run dynamics relating to the internal structures of

13. For a survey of the several classes of molecular and intermolecular forces, and their dissociation energies, see Setlow and Pollard, *Molecular Biophysics,* chapter 6. The energies of typical covalent bonds are of the order of 80–100 k cal/mole, of the hydrogen bonds, 10 k cal/mole. Ionic bonds generally lie between these two levels; the bonds due to van der Waals forces are lower in energy.

14. Typical wave numbers for vibrations associated with various systems (the wave number is the reciprocal of wave length, hence proportional to frequency):
Steel wire under tension—10^{-10} to 10^{-9} cm^{-1}
Molecular rotations—10^0 to 10^2 cm^{-1}
Molecular vibrations—10^2 to 10^3 cm^{-1}
Planetary electrons—10^4 to 10^5 cm^{-1}
Nuclear rotations—10^9 to 10^{10} cm^{-1}
Nuclear surface vibrations—10^{11} to 10^{12} cm^{-1}

the subsystems and the long-run dynamics to the interactions of these subsystems.

A number of the important approximations employed in physics depend for their validity on the near decomposability of the systems studied. The theory of the thermodynamics of irreversible processes, for example, requires the assumption of macroscopic disequilibrium but microscopic equilibrium, exactly the situation described in our heat-exchange example.[15] Similarly computations in quantum mechanics are often handled by treating weak interactions as producing perturbations on a system of strong interactions.

Some Observations on Hierarchic Span

To understand why the span of hierarchies is sometimes very broad—as in crystals—and sometimes narrow, we need to examine more detail of the interactions. In general the critical consideration is the extent to which interaction between two (or a few) subsystems excludes interaction of these subsystems with the others. Let us examine first some physical examples.

Consider a gas of identical molecules, each of which can form covalent bonds in certain ways with others. Let us suppose that we can associate with each atom a specific number of bonds that it is capable of maintaining simultaneously. (This number is obviously related to the number we usually call its valence.) Now suppose that two atoms join and that we can also associate with the combination a specific number of external bonds it is capable of maintaining. If this number is the same as the number associated with the individual atoms, the bonding process can go on indefinitely—the atoms can form crystals or polymers of indefinite extent. If the number of bonds of which the composite is capable is less than the number associated with each of the parts, then the process of agglomeration must come to a halt.

We need only mention some elementary examples. Ordinary gases show no tendency to agglomerate, because the multiple bonding of atoms "uses up" their capacity to interact. While each oxygen atom has a valence of two, the O_2 molecules have a zero valence. Contrariwise, indefi-

15. S. R. de Groot, *Thermodynamics of Irreversible Processes* (New York: Interscience Publishers, 1951), pp. 11–12.

nite chains of single-bonded carbon atoms can be built up, because a chain of any number of such atoms, each with two side groups, has a valence of exactly two.

Now what happens if we have a system of elements that possess both strong and weak interaction capacities and whose strong bonds are exhaustible through combination? Subsystems will form, until all the capacity for strong interaction is utilized in their construction. Then these subsystems will be linked by the weaker second-order bonds into larger systems. For example, a water molecule has essentially a valence of zero—all the potential covalent bonds are fully occupied by the interaction of hydrogen and oxygen molecules. But the geometry of the molecule creates an electric dipole that permits weak interaction between the water and salts dissolved in it—whence such phenomena as its electrolytic conductivity.[16]

Similarly it has been observed that, although electrical forces are much stronger than gravitational forces, the latter are far more important than the former for systems on an astronomical scale. The explanation of course is that the electrical forces, being bipolar, are all "used up" in the linkages of the smaller subsystems and that significant net balances of positive or negative charges are not generally found in regions of macroscopic size.

In social as in physical systems there are generally limits on the simultaneous interaction of large numbers of subsystems. In the social case these limits are related to the fact that a human being is more nearly a serial than a parallel information-processing system. He or she can carry on only one conversation at a time, and although this does not limit the size of the audience to which a mass communication can be addressed, it does limit the number of people simultaneously involved in most other forms of social interaction. Apart from requirements of direct interactions, most roles impose tasks and responsibilities that are time consuming. One cannot, for example, enact the role of "friend" with large numbers of other people.

It is probably true that in social as in physical systems the higher-frequency dynamics are associated with the subsystems and the lower-

16. See, for example, L. Pauling, *General Chemistry* (San Francisco: W. H. Freeman, 2nd ed., 1953), chapter 15.

frequency dynamics with the larger systems. It is generally believed, for example, that the relevant planning horizon of executives is longer, the higher their location in the organizational hierarchy. It is probably also true that both the average duration of an interaction between executives and the average interval between interactions are greater at higher than lower levels.

Summary: Near Decomposability

We have seen that hierarchies have the property of near decomposability. Intracomponent linkages are generally stronger than intercomponent linkages. This fact has the effect of separating the high-frequency dynamics of a hierarchy—involving the internal structure of the components—from the low-frequency dynamics—involving interaction among components. We shall turn next to some important consequences of this separation for the description and comprehension of complex systems.

Biological Evolution Revisited

Having examined the properties of nearly-decomposable systems, we can now complete our discussion of the evolution of multi-cellular organisms through specialization of tissues and organs. An organ performs a specific set of functions, each usually requiring continual interaction among its component parts (a sequence of chemical reactions, say, each step employing a particular enzyme for its execution). It draws raw materials from other parts of the organism and delivers products to other parts, but these input and output processes depend only in an aggregate way on what is occurring within each specific organ. Like a business firm in an economic market, each organ can perform its functions in blissful ignorance of the detail of activity in other organs, with which it is connected by the digestive, circulatory, and excretory systems and other transport channels.

Expressing the matter slightly differently, changes within an organ affect the other parts of the organism mainly by changing the relation between the quantities of outputs they produce and the inputs they require (that is, their overall efficiency). Thus, biological organisms are nearly-decomposable: the interactions *within* units at any level are rapid and intense in comparison with the interactions *between* units at the same level. Inventories of various substances, held in the circulatory system or in special tissues, slow down and buffer effects of each unit on the others. In the short run, single units (e.g., single organs) operate nearly independently of the detail of operation of the other units.

Within the Darwinian framework of natural selection there is no way in which the fitness (efficiency) of individual tissues or organs can be separately evaluated; fitness is measured by the number of offspring of the entire organism. Evolution is like a complex experiment, with fitness as the sole dependent variable, and the structures of the individual genes as independent variables. The goal of the process is to compare the contribution to total fitness of alternative forms (alleles) of each gene—and of combinations of these alternatives for sets of genes.

If, in fact, the fitness of a particular gene depended on which alleles of all the other genes it was combined with, the combinatorics, involving tens of thousands of genes in complex organisms, would be staggering, and the problem of measuring the contribution of a particular allele to fitness would be overwhelming.[17]

With near-decomposability, we can assume that the *relative* efficiency of two different designs for the same organ (e.g., two different gene sequences with the same function) is approximately independent of which variants of other organs are present in the organism. The total fitness is essentially an additive measure of the nearly independent contributions of the individual organs. Essentially, we obtain the advantages of the clicking safe: the "correct" setting of each dial (the genes governing one organ's processes) can be determined independently of how the other dials are currently set. The search is for effective sets of organs instead of effective sets of individual genes.

Enough is known today about the architecture of the genome to be reasonably certain that it has a hierarchical control structure mapping reasonably closely to the hierarchy of processes in the organism.[18] Of

17. With only two alleles for each of N genes, 2^N alternatives would have to be evaluated by selection. This is equivalent, in the watchmaker metaphor, to assembling 2^N parts without interruption. For an organism with even a thousand genes, say, change by natural selection would be extremely slow, even on a geological scale.

18. F. Jacob and J. Monod, "Genetic Regulatory Mechanisms in the Synthesis of Proteins," *Molecular Biology,* 3(1961):318–56.

course, this is a gross simplification of the total picture in any actual organism. In addition to the genes that operate in particular organs (turned on and off by control genes), there are also the genes that determine the more general metabolic processes that are found within all the cells. But these common intra-cellular processes are at the cell level of the hierarchy, below the level of tissues and organs, and again the corresponding genes can be supposed to operate nearly independently of those that control specialized processes in specific organs.[19]

The Description of Complexity

If you ask a person to draw a complex object—such as a human face—he will almost always proceed in a hierarchic fashion.[20] First he will outline the face. Then he will add or insert features: eyes, nose, mouth, ears, hair. If asked to elaborate, he will begin to develop details for each of the features—pupils, eyelids, lashes for the eyes, and so on—until he reaches the limits of his anatomical knowledge. His information about the object is arranged hierarchically in memory, like a topical outline.

When information is put in outline form, it is easy to include information about the relations among the major parts and information about the internal relations of parts in each of the suboutlines. Detailed information about the relations of subparts belonging to different parts has no place in the outline and is likely to be lost. The loss of such information and the preservation mainly of information about hierarchic order is a salient characteristic that distinguishes the drawings of a child or someone untrained in representation from the drawing of a trained artist. (I am speaking of an artist who is striving for representation.)

19. How hierarchical architectures of these kinds can be introduced into the genetic algorithms discussed in chapter 7, in order to speed up their rates of learning or evolution, is discussed by John H. Holland in *Adaptation in Natural and Artificial Systems* (Ann Arbor, MI: The University of Michigan Press, 1975). See especially pp. 167–168 and 152–153.

20. George A. Miller has collected protocols from subjects who were given the task of drawing faces and finds that they behave in the manner described here (private communication). See also E. H. Gombrich, *Art and Illusion* (New York: Pantheon Books, 1960), pp. 291–296.

Near Decomposability and Comprehensibility
From our discussion of the dynamic properties of nearly decomposable systems, we have seen that comparatively little information is lost by representing them as hierarchies. Subparts belonging to different parts only interact in an aggregative fashion—the detail of their interaction can be ignored. In studying the interaction of two large molecules, generally we do not need to consider in detail the interactions of nuclei of the atoms belonging to the one molecule with the nuclei of the atoms belonging to the other. In studying the interaction of two nations, we do not need to study in detail the interactions of each citizen of the first with each citizen of the second.

The fact then that many complex systems have a nearly decomposable, hierarchic structure is a major facilitating factor enabling us to understand, describe, and even "see" such systems and their parts. Or perhaps the proposition should be put the other way round. If there are important systems in the world that are complex without being hierarchic, they may to a considerable extent escape our observation and understanding. Analysis of their behavior would involve such detailed knowledge and calculation of the interactions of their elementary parts that it would be beyond our capacities of memory or computation.[21]

21. I believe the fallacy in the central thesis of W. M. Elsasser's *The Physical Foundation of Biology,* mentioned earlier, lies in his ignoring the simplification in description of complex systems that derives from their hierarchic structure. Thus (p. 155):

If we now apply similar arguments to the coupling of enzymatic reactions with the substratum of protein molecules, we see that over a sufficient period of time, the information corresponding to the structural details of these molecules will be communicated to the dynamics of the cell, to higher levels of organization as it were, and may influence such dynamics. While this reasoning is only qualitative, it lends credence to the assumption that in the living organism, unlike the inorganic crystal, the effects of microscopic structure cannot be simply averaged out; as time goes on this influence will pervade the behavior of the cell "at all levels."

But from our discussion of near decomposability it would appear that those aspects of microstructure that control the slow developmental aspects of organismic dynamics can be separated out from the aspects that control the more rapid cellular metabolic processes. For this reason we should not despair of unraveling the web of causes. See also J. R. Platt's review of Elsasser's book in *Perspectives in Biology and Medicine,* 2(1959):243–245.

I shall not try to settle which is chicken and which is egg: whether we are able to understand the world because it is hierarchic or whether it appears hierarchic because those aspects of it which are not elude our understanding and observation. I have already given some reasons for supposing that the former is at least half the truth—that evolving complexity would tend to be hierarchic—but it may not be the whole truth.

Simple Descriptions of Complex Systems

One might suppose that the description of a complex system would itself be a complex structure of symbols—and indeed it may be just that. But there is no conservation law that requires that the description be as cumbersome as the object described. A trivial example will show how a system can be described economically. Suppose the system is a two-dimensional array like this:

$$
\begin{array}{cccccccc}
A & B & M & N & R & S & H & I \\
C & D & O & P & T & U & J & K \\
M & N & A & B & H & I & R & S \\
O & P & C & D & J & K & T & U \\
R & S & H & I & A & B & M & N \\
T & U & J & K & C & D & O & P \\
H & I & R & S & M & N & A & B \\
J & K & T & U & O & P & C & D \\
\end{array}
$$

Let us call the array $\begin{vmatrix} AB \\ CD \end{vmatrix}$ a, the array $\begin{vmatrix} MN \\ OP \end{vmatrix}$ m, the array $\begin{vmatrix} RS \\ TU \end{vmatrix}$ r, and the array $\begin{vmatrix} HI \\ JK \end{vmatrix}$ h. Let us call the array $\begin{vmatrix} am \\ ma \end{vmatrix}$ w, and the array $\begin{vmatrix} rh \\ hr \end{vmatrix}$ x. Then the entire array is simply $\begin{vmatrix} wx \\ xw \end{vmatrix}$. While the original structure consisted of 64 symbols, it requires only 35 to write down its description:

$$
S = \frac{wx}{xw}
$$

$$
w = \frac{am}{ma} \qquad\qquad x = \frac{rh}{hr}
$$

$$
a = \frac{AB}{CD} \qquad m = \frac{MN}{OP} \qquad r = \frac{RS}{TU} \qquad h = \frac{HI}{JK}
$$

We achieve the abbreviation by making use of the redundancy in the original structure. Since the pattern $\begin{vmatrix} AB \\ CD \end{vmatrix}$, for example, occurs four times in the total pattern, it is economical to represent it by the single symbol, a.

If a complex structure is completely unredundant—if no aspect of its structure can be inferred from any other—then it is its own simplest description. We can exhibit it, but we cannot describe it by a simpler structure. The hierarchic structures we have been discussing have a high degree of redundancy, hence can often be described in economical terms. The redundancy takes a number of forms, of which I shall mention three:

1. Hierarchic systems are usually composed of only a few different kinds of subsystems in various combinations and arrangements. A familiar example is the proteins, their multitudinous variety arising from arrangements of only twenty different amino acids. Similarly the ninety-odd elements provide all the kinds of building blocks needed for an infinite variety of molecules. Hence we can construct our description from a restricted alphabet of elementary terms corresponding to the basic set of elementary subsystems from which the complex system is generated.

2. Hierarchic systems are, as we have seen, often nearly decomposable. Hence only aggregative properties of their parts enter into the description of the interactions of those parts. A generalization of the notion of near decomposability might be called the "empty world hypothesis"—most things are only weakly connected with most other things; for a tolerable description of reality only a tiny fraction of all possible interactions needs to be taken into account. By adopting a descriptive language that allows the absence of something to go unmentioned, a nearly empty world can be described quite concisely. Mother Hubbard did not have to check off the list of possible contents to say that her cupboard was bare.

3. By appropriate "recoding," the redundancy that is present but unobvious in the structure of a complex system can often be made patent. The commonest recoding of descriptions of dynamic systems consists in replacing a description of the time path with a description of a differential law that generates that path. The simplicity resides in a constant relation between the state of the system at any given time and the state of the system a short time later. Thus the structure of the sequence 1 3 5 7 9 11 . . . is most simply expressed by observing that each member is obtained by adding 2 to the previous one. But this is the sequence that Galileo found to describe the velocity at the end of successive time intervals of a ball rolling down an inclined plane.

It is a familiar proposition that the task of science is to make use of the world's redundancy to describe that world simply. I shall not pursue the general methodological point here, but I shall instead take a closer look at two main types of description that seem to be available to us in seeking an understanding of complex systems. I shall call these *state description* and *process description,* respectively.

State Descriptions and Process Descriptions

"A circle is the locus of all points equidistant from a given point." "To construct a circle, rotate a compass with one arm fixed until the other arm has returned to its starting point." It is implicit in Euclid that if you carry out the process specified in the second sentence, you will produce an object that satisfies the definition of the first. The first sentence is a state description of a circle; the second, a process description.

These two modes of apprehending structures are the warp and weft of our experience. Pictures, blueprints, most diagrams, and chemical structural formulas are state descriptions. Recipes, differential equations, and equations for chemical reactions are process descriptions. The former characterize the world as sensed; they provide the criteria for identifying objects, often by modeling the objects themselves. The latter characterize the world as acted upon; they provide the means for producing or generating objects having the desired characteristics.

The distinction between the world as sensed and the world as acted upon defines the basic condition for the survival of adaptive organisms. The organism must develop correlations between goals in the sensed world and actions in the world of process. When they are made conscious and verbalized, these correlations correspond to what we usually call means-ends analysis. Given a desired state of affairs and an existing state of affairs, the task of an adaptive organism is to find the difference between these two states and then to find the correlating process that will erase the difference.[22]

Thus problem solving requires continual translation between the state and process descriptions of the same complex reality. Plato, in the *Meno,*

22. See H. A. Simon and A. Newell, "Simulation of Human Thinking," in M. Greenberger (ed.), *Management and the Computer of the Future* (New York: Wiley, 1962), pp. 95–114, esp. pp. 110 ff.

argued that all learning is remembering. He could not otherwise explain how we can discover or recognize the answer to a problem unless we already know the answer.[23] Our dual relation to the world is the source and solution of the paradox. We pose a problem by giving the state description of the solution. The task is to discover a sequence of processes that will produce the goal state from an initial state. Translation from the process description to the state description enables us to recognize when we have succeeded. The solution is genuinely new to us—and we do not need Plato's theory of remembering to explain how we recognize it.

There is now a growing body of evidence that the activity called human problem solving is basically a form of means-ends analysis that aims at discovering a process description of the path that leads to a desired goal. The general paradigm is: Given a blueprint, to find the corresponding recipe. Much of the activity of science is an application of that paradigm: Given the description of some natural phenomena, to find the differential equations for processes that will produce the phenomena.

The Description of Complexity in Self-Reproducing Systems

The problem of finding relatively simple descriptions for complex systems is of interest not only for an understanding of human knowledge of the world but also for an explanation of how a complex system can reproduce itself. In my discussion of the evolution of complex systems, I touched only briefly on the role of self-reproduction.

Atoms of high atomic weight and complex inorganic molecules are witnesses to the fact that the evolution of complexity does not imply self-reproduction. If evolution of complexity from simplicity is sufficiently probable, it will occur repeatedly; the statistical equilibrium of the system will find a large fraction of the elementary particles participating in complex systems.

If, however, the existence of a particular complex form increased the probability of the creation of another form just like it, the equilibrium between complexes and components could be greatly altered in favor of the former. If we have a description of an object that is sufficiently clear

23. *The Works of Plato*, B. Jowett, translator (New York: Dial Press, 1936), vol. 3, pp. 26–35. See H. A. Simon, "Bradie on Polanyi on the Meno Paradox," *Philosophy of Science*, 43(1976):147–150.

and complete, we can reproduce the object from the description. Whatever the exact mechanism of reproduction, the description provides us with the necessary information.

Now we have seen that the descriptions of complex systems can take many forms. In particular we can have state descriptions, or we can have process descriptions—blueprints or recipes. Reproductive processes could be built around either of these sources of information. Perhaps the simplest possibility is for the complex system to serve as a description of itself—a template on which a copy can be formed. One of the most plausible current theories, for example, of the reproduction of deoxyribonucleic acid (DNA) proposes that a DNA molecule, in the form of a double helix of matching parts (each essentially a "negative" of the other), unwinds to allow each half of the helix to serve as a template on which a new matching half can form.

On the other hand, our current knowledge of how DNA controls the metabolism of the organism suggests that reproduction by template is only one of the processes involved. According to the prevailing theory, DNA serves as a template both for itself and for the related substance ribonucleic acid (RNA). RNA in turn serves as a template for protein. But proteins—according to current knowledge—guide the organism's metabolism not by the template method but by serving as catalysts to govern reaction rates in the cell. While RNA is a blueprint for protein, protein is a recipe for metabolism.[24]

Ontogeny Recapitulates Phylogeny

The DNA in the chromosomes of an organism contains some, and perhaps most, of the information that is needed to determine its development and activity. We have seen that, if current theories are even approximately correct, the information is recorded not as a state description of the organism but as a series of "instructions" for the construction and maintenance of the organism from nutrient materials. I have already used the

24. C. B. Anfinsen, *The Molecular Basis of Evolution* (New York: Wiley, 1959), chapters 3 and 10, will qualify this sketchy, oversimplified account. For an imaginative discussion of some mechanisms of process description that could govern molecular structure, see H. H. Pattee, "On the Origin of Macromolecular Sequences," *Biophysical Journal,* 1(1961):683–710.

metaphor of a recipe; I could equally well compare it with a computer program, which is also a sequence of instructions governing the construction of symbolic structures. Let me spin out some of the consequences of the latter comparison.

If genetic material is a program—viewed in its relation to the organism—it is a program with special and peculiar properties. First, it is a self-reproducing program; we have already considered its possible copying mechanism. Second, it is a program that has developed by Darwinian evolution. On the basis of our watchmaker's argument, we may assert that many of its ancestors were also viable programs—programs for the subassemblies.

Are there any other conjectures we can make about the structure of this program? There is a well-known generalization in biology that is verbally so neat that we would be reluctant to give it up even if the facts did not support it: ontogeny recapitulates phylogeny. The individual organism in its development goes through stages that resemble some of its ancestral forms. The fact that the human embryo develops gill bars and then modifies them for other purposes is a familiar particular belonging to the generalization. Biologists today like to emphasize the qualifications of the principle—that ontogeny recapitulates only the grossest aspects of phylogeny, and these only crudely. These qualifications should not make us lose sight of the fact that the generalization does hold in rough approximation—it does summarize a very significant set of facts about the organism's development. How can we interpret these facts?

One way to solve a complex problem is to reduce it to a problem previously solved—to show what steps lead from the earlier solution to a solution of the new problem. If around the turn of the century we wanted to instruct a workman to make an automobile, perhaps the simplest way would have been to tell him how to modify a wagon by removing the singletree and adding a motor and transmission. Similarly a genetic program could be altered in the course of evolution by adding new processes that would modify a simpler form into a more complex one—to construct a gastrula, take a blastula and alter it!

The genetic description of a single cell may therefore take a quite different form from the genetic description that assembles cells into a multicelled organism. Multiplication by cell division would require as a minimum a state description (the DNA, say), and a simple "interpretive process"—to use the term from computer language—that copies this description as a part of the larger copying process of cell division. But such a mechanism clearly would not suffice for the differentiation of cells in development. It appears more natural to conceptualize that mechanism as based on a process description and a somewhat more complex interpretive process that produces the adult organism in a sequence of stages, each new stage in development representing the effect of an operator upon the previous one.

It is harder to conceptualize the interrelation of these two descriptions. Interrelated they must be, for enough has been learned of gene-enzyme mechanisms to show that these play a major role in development as in cell metabolism. The single clue we obtain from our earlier discussion is that the description may itself be hierarchical, or nearly decomposable, in structure, the lower levels governing the fast, "high-frequency" dynamics of the individual cell and the higher-level interactions governing the slow, "low-frequency" dynamics of the developing multicellular organism.

There is a rapidly growing body of evidence that the genetic program is organized in this way.[25] To the extent that we can differentiate the genetic information that governs cell metabolism from the genetic information that governs the development of differentiated cells in the multicellular organization, we simplify enormously—as we have already seen—our

---

25. For extensive discussion of these matters, see J. D. Watson, *op. cit.*, especially chapters 8 and 14. For a review of some of the early evidence, see P. E. Hartman, "Transduction: A Comparative Review," in W. D. McElroy and B. Glass (eds.), *The Chemical Basis of Heredity* (Baltimore: Johns Hopkins Press, 1957), pp. 442–454. Evidence for differential activity of genes in different tissues and at different stages of development is discussed by J. G. Gall, "Chromosomal Differentiation," in W. D. McElroy and B. Glass (eds.), *The Chemical Basis of Development* (Baltimore: Johns Hopkins Press, 1958), pp. 103–135. Finally, a model very like that proposed here has been independently, and far more fully, outlined by J. R. Platt, "A 'Book Model' of Genetic Information Transfer in Cells and Tissues," in M. Kasha and B. Pullman (eds.), *Horizons in Biochemistry* (New York: Academic Press, 1962), pp. 167–187. Of course this kind of mechanism is not the only one in which development could be controlled by a process description. Induction, in the form envisaged in Spemann's organizer theory, is based on a process description in which metabolites in already formed tissue control the next stages of development.

task of theoretical description. But I have perhaps pressed this speculation far enough.

The generalization that we might expect ontogeny partially to recapitulate phylogeny in evolving systems whose descriptions are stored in a process language has applications outside the realm of biology. It can be applied as readily, for example, to the transmission of knowledge in the educational process. In most subjects, particularly in the rapidly advancing sciences, the progress from elementary to advanced courses is to a considerable extent a progress through the conceptual history of the science itself. Fortunately the recapitulation is seldom literal—any more than it is in the biological case. We do not teach the phlogiston theory in chemistry in order later to correct it. (I am not sure I could not cite examples in other subjects where we do exactly that.) But curriculum revisions that rid us of the accumulations of the past are infrequent and painful. Nor are they always desirable—partial recapitulation may, in many instances, provide the most expeditious route to advanced knowledge.

Summary: The Description of Complexity

How complex or simple a structure is depends critically upon the way in which we describe it. Most of the complex structures found in the world are enormously redundant, and we can use this redundancy to simplify their description. But to use it, to achieve the simplification, we must find the right representation.

The notion of substituting a process description for a state description of nature has played a central role in the development of modern science. Dynamic laws, expressed in the form of systems of differential or difference equations, have in a large number of cases provided the clue for the simple description of the complex. In the preceding paragraphs I have tried to show that this characteristic of scientific inquiry is not accidental or superficial. The correlation between state description and process description is basic to the functioning of any adaptive organism, to its capacity for acting purposefully upon its environment. Our present-day understanding of genetic mechanisms suggests that even in describing itself the multicellular organism finds a process description—a genetically encoded program—to be the parsimonious and useful representation.

Conclusion

Our speculations have carried us over a rather alarming array of topics, but that is the price we must pay if we wish to seek properties common to many sorts of complex systems. My thesis has been that one path to the construction of a nontrivial theory of complex systems is by way of a theory of hierarchy. Empirically a large proportion of the complex systems we observe in nature exhibit hierarchic structure. On theoretical grounds we could expect complex systems to be hierarchies in a world in which complexity had to evolve from simplicity. In their dynamics hierarchies have a property, near decomposability, that greatly simplifies their behavior. Near decomposability also simplifies the description of a complex system and makes it easier to understand how the information needed for the development or reproduction of the system can be stored in reasonable compass.

In science and engineering the study of "systems" is an increasingly popular activity. Its popularity is more a response to a pressing need for synthesizing and analyzing complexity than it is to any large development of a body of knowledge and technique for dealing with complexity. If this popularity is to be more than a fad, necessity will have to mother invention and provide substance to go with the name. The explorations reviewed here represent one particular direction of search for such substance.

Name Index

Subject Index